Employee-Driven Innovation

Employee-Driven Innovation
A New Approach

Edited by

Steen Høyrup
Aarhus University, Denmark

Maria Bonnafous-Boucher
Novancia, Paris, France

Cathrine Hasse
Aarhus University, Denmark

Maja Lotz
Copenhagen Business School, Denmark

Kirsten Møller
Aarhus University, Denmark

GUELPH HUMBER LIBRARY
205 Humber College Blvd
Toronto, ON   M9W 5L7

Selection and editorial content © Steen Høyrup, Maria Bonnafous-Boucher, Cathrine Hasse, Maja Lotz, Kirsten Møller 2012
Individual chapters © the contributors 2012
Foreword © Tara Fenwick 2012

All rights reserved. No reproduction, copy or transmission of this publication may be made without written permission.

No portion of this publication may be reproduced, copied or transmitted save with written permission or in accordance with the provisions of the Copyright, Designs and Patents Act 1988, or under the terms of any licence permitting limited copying issued by the Copyright Licensing Agency, Saffron House, 6–10 Kirby Street, London EC1N 8TS.

Any person who does any unauthorized act in relation to this publication may be liable to criminal prosecution and civil claims for damages.

The authors have asserted their rights to be identified as the authors of this work in accordance with the Copyright, Designs and Patents Act 1988.

First published 2012 by
PALGRAVE MACMILLAN

Palgrave Macmillan in the UK is an imprint of Macmillan Publishers Limited, registered in England, company number 785998, of Houndmills, Basingstoke, Hampshire RG21 6XS.

Palgrave Macmillan in the US is a division of St Martin's Press LLC, 175 Fifth Avenue, New York, NY 10010.

Palgrave Macmillan is the global academic imprint of the above companies and has companies and representatives throughout the world.

Palgrave® and Macmillan® are registered trademarks in the United States, the United Kingdom, Europe and other countries

ISBN: 978–0–230–27862–2

This book is printed on paper suitable for recycling and made from fully managed and sustained forest sources. Logging, pulping and manufacturing processes are expected to conform to the environmental regulations of the country of origin.

A catalogue record for this book is available from the British Library.

Library of Congress Cataloging-in-Publication Data

Employee-driven innovation : a new approach / edited by Steen Høyrup
 ... [et al.].
   p. cm.
 Includes bibliographical references.
 ISBN 978–0–230–27862–2
   1. Technological innovations – Employee participation.
 I. Høyrup, Steen.
HD45.E367 2012
658.3'14—dc23   2012011154

10 9 8 7 6 5 4 3 2 1
21 20 19 18 17 16 15 14 13 12

Printed and bound in the United States of America
by Edwards Brothers Malloy, Inc.

Contents

List of Tables vii

List of Figures viii

Foreword by Tara Fenwick ix

Acknowledgements xi

Notes on Contributors xii

List of Abbreviations xvii

Part I The Nature of Employee-Driven Innovation

1 Employee-Driven Innovation: A New Phenomenon, Concept and Mode of Innovation 3
   Steen Høyrup

2 Employee-Driven Innovation: Operating in a Chiaroscuro 34
   Ann-Charlotte Teglborg, Renaud Redien-Collot, Maria Bonnafous-Boucher and Céline Viala

3 In Search of Best Practices for Employee-Driven Innovation: Experiences from Norwegian Work Life 57
   Tone Merethe Aasen, Oscar Amundsen, Leif Jarle Gressgård and Kåre Hansen

Part II Employee-Driven Innovation in the Workplace Mediated through Employees' Learning

4 Creating Work: Employee-Driven Innovation through Work Practice Reconstruction 77
   Oriana Milani Price, David Boud and Hermine Scheeres

5 Explaining Innovation at Work: A Socio-Personal Account 92
   Stephen Billett

6 Innovation Competency – An Essential Organizational Asset 108
   Lotte Darsø

7 Employee-Driven Innovation and Practice-Based Learning in Organizational Cultures 127
   Ulrik Brandi and Cathrine Hasse

8 Employee-Driven Innovation Amongst 'Routine' Employees in the UK: The Role of Organizational 'Strategies' and Individual 'Tactics' 149
Edmund Waite, Karen Evans and Natasha Kersh

Part III Employee-Driven Innovation Unfolded in Global Networks and Complex Systems

9 Moving Organizations toward Employee-Driven Innovation (EDI) in Work Practices and on a Global Scale: Possibilities and Challenges 167
Maja Lotz and Peer Hull Kristensen

10 Exploring the Employee-Driven Innovation Concept by Comparing 'Innovation Capability Management' Among German and Chinese Firms 185
Werner Fees and Amir H. Taherizadeh

11 Privileged Yet Restricted? Employee-Driven Innovation and Learning in Three R&D Communities 211
Tea Lempiälä and Sari Yli-Kauhaluoma

12 Employee-Driven Innovation and Industrial Relations 230
Stan De Spiegelaere and Guy Van Gyes

Index 247

Tables

2.1	Favi and Solvay-Tavaux satisfy the three variables of entrepreneurial orientation	43
3.1	Case enterprises	61
10.1	Summary of the data collection	194
10.2	Descriptive data	195
10.3	Average performance on each innovation dimension within German and Chinese firms	196
10.4	Average total innovation capability in German and Chinese firms	197
10.5	Detailed comparative analysis of five key dimensions of successful innovation management	199
11.1	Three R&D communities	216
11.2	Results of cross-case comparison	218
12.1	Works councils and innovation	234
12.2	Unions and innovation	237
12.3	Possible effects of employee participation on innovation processes	238

Figures

1.1	EDI in an organizational and global context	12
3.1	Interrelated elements of EDI	68
6.1	The Diamond of Innovation	111
6.2	Preject–project	115
6.3	Leadership roles	115
6.4	The Diamond of Innovation linked to four types of knowing	116
6.5	Framework for innovation competency	118
10.1	Key areas for successful innovation management	192
10.2	Average performance on each innovation dimension within German and Chinese firms	197

Foreword

Innovation is a powerful and pervasive discourse in workplaces these days. In both private and public sector organizations, innovation is promoted as the critical engine of growth, or perhaps just survival, in hyper-competitive global capitalism. Policy documents stress innovation as a critical skill in the so-called knowledge economy. The most desirable employees are those who are innovative. The most important organizational activities are those dedicated to generating, capturing, institutionalizing and marketing innovation. With so much at stake, hierarchies have flourished in the divisions that are taken to exist between those elite designers valued as the *innovators* and those others who so often are assumed to be the *implementers* – that is, those who engineer, maintain and imaginatively negotiate a mass of everyday problem-solving to make innovations actually work.

But just what is 'innovation', and whom does it serve? In critical circles, important questions have been raised about this discourse for some time. What counts as innovative knowledge tends to be that which is considered novel, solves practical problems, and commands market value. But, of course, all this is determined by very particular interests that recognize and value most those ideas that yield maximum exchange value in networks of production. What tends to be overlooked are other forms of innovation. These include those everyday improvisations that workers generate all the time: small modifications to work processes, inventive solutions that emerge in practice, everyday openings for creative expression that workers somehow find and exploit, and even innovative approaches to well-being. What also tend to become obscured in organizational discussions of innovation are both creative *challenges* to the hegemonic discourses of profit, expansion and productivity driving the current global economy, and creative *alternatives* to received categories and processes.

The authors in this collection have adopted a unique vantage point from which to consider issues of innovation: the employee who 'drives' innovation. What constitutes innovation, from the employees' perspectives? What do they value as innovation? The cases reported here show us glimpses of workers' everyday activity, where innovation processes unfold in the very activities of practice. That is, the divisions between innovation and implementation, or between designing and using, are much more enmeshed than conventional portrayals of innovation have assumed. Further, these cases provide insights into workers' motivations for innovating. While some employee-driven innovations may be subversive, most tend to be genuinely oriented towards seeding more productive, effective and sustainable work processes. Indeed, many employees generate innovative work-arounds simply to make things work that wouldn't otherwise, or to stretch resources when they don't have the tools they need. In this sense, employee-driven innovation can draw necessary attention to problematic work conditions.

Broader questions raised in the discussions here about employee-driven innovation pertain to ownership. In a knowledge economy, innovative ideas are important capital. To what extent, and in what circumstances, do the employee and the organization respectively deserve recognition and material benefit from employees' innovative activity? Laws governing patents and copyrights are struggling to keep up with the issues posed by the co-production of ideas, products and processes involving employees as well as users and designers. The viral flourishing of virtual innovations, and the growing internet ethos of sharing creative material and ideas in a logic of open access, also raise interesting questions about the terms of exchange and dissemination of innovative property generated by employees.

At the broadest level, these chapters refocus the discussion about the purposes and nature of innovation in work organizations. What counts as good innovation, and when is innovation problematic? Whose assessment determines which innovations become amplified and extended, which are rewarded, and which are stifled? How do different organizational stakeholders – clients, trade unions, government, global markets, financiers, regulatory agencies, and so on – affect the processes of employee-driven innovation? Some authors draw attention to different forms of innovation, and suggest the need for new vocabularies with which to analyse and represent innovative activity. Some raise central questions about the 'goods' that are taken to be inherent in innovative activity, innovative products, and the current prevalence of the innovation discourse in work, policy and learning.

Finally, and perhaps most importantly, this collection causes the reader to question whose interests are served through this pervasive press for innovation in work, and the extent to which opportunities for employee-driven innovation benefit the workers doing the driving.

Tara Fenwick
University of Stirling, UK
October 2011

Acknowledgements

Thank you to J. Tidd, J. Bessant and K. Pavitt for permission to use the Figure: *Key areas for successful innovation management* (Figure 10.1, page 192).

Contributors

Tone Merethe Aasen is a research manager at NTNU Social Research AS, Norway. She holds an MSc in Biophysics and Medical Engineering from Norwegian University of Science and Technology (NTNU), an MMa from BI Norwegian School of Management and a PhD in Organizational Innovation Processes from NTNU. Her main research interests include innovation processes and innovation management, organizational complexity and business performance.

Oscar Amundsen is a researcher in the Department of Adult Learning and Counselling, Norwegian University of Science and Technology in Trondheim, Norway. His Master's is in Language and Literature Studies and includes disciplines from social sciences, humanities and technology studies. His PhD is in Organizational Studies, and his main research interests are change, innovation, development and culture in organizations.

Stephen Billett is Professor of Adult and Vocational Education in the School of Education and Professional Studies at Griffith University, Brisbane, Australia. Stephen has worked as a vocational educator, educational administrator, teacher educator, professional development practitioner and policy developer within the Australian vocational education system and as a teacher and researcher at Griffith University. Since 1992, he has researched learning through and for work and has published widely in the fields of vocational learning, workplace learning and conceptual accounts of learning for vocational purposes.

Maria Bonnafous-Boucher is a researcher and Dean of Research at Novancia Business School, Paris. She is a Co-Scientific Director of the Chair of Research in Entrepreneurship and Innovation, Paris Chamber of Commerce and Industry, France. She holds a PhD in Strategy and Organizational Studies and a Master's in Philosophy, majoring in Politics and Morals. Her main research interests include institutional entrepreneurship (clusters policy and governance), stakeholder theory (governance), business ethics and organizational studies.

David Boud is Professor of Adult Education at the University of Technology, Sydney. His research interests are in the areas of learning in workplaces, continuing professional education and assessment for learning in higher and professional education.

Ulrik Brandi is Associate Professor in the Department of Education, Aarhus University. Ulrik holds a Master's in Educational Studies and Philosophy and a PhD in Organizational Learning. His research interests include organizational learning and change, knowledge sharing and innovation.

Lotte Darsø is Associate Professor of Innovation in the Department of Education, Aarhus University, Denmark. Lotte holds a Master's in Social Psychology from Copenhagen University and a PhD from Copenhagen Business School. Her main research interests include creativity, innovation processes, innovation pedagogies and creative approaches to organizational innovation and transformation.

Stan De Spiegelaere is a researcher at HIVA (Research Institute for Work and Society), Catholic University of Leuven (K.U. Leuven). He is a graduate in Political Science (2007) and Labour Sciences (2010). His doctoral research covers the themes of labour regulation, industrial relations and innovative work behaviour of employees.

Karen Evans is Professor and Chair in Education (Lifelong Learning) at the Institute of Education, University of London. She holds a PhD from the University of Surrey and is an academician of the Academy of Social Sciences. She has authored numerous books on learning in working life, including *Learning, Work and Social Responsibility* (2009) and *Improving Literacy at Work* (2011). She is co-editor of the *Sage Handbook of Workplace Learning* (2011) and the *Second International Handbook of Lifelong Learning* (2011).

Werner Fees is Professor of Management at Georg Simon Ohm University of Applied Sciences, Nuremberg, Germany. He holds a PhD from Friedrich-Alexander-Universität Erlangen-Nuremberg (FAU), Germany. His main research interests include innovation and strategic management, Chinese and German management theories and international business.

Tara Fenwick is Professor of Professional Education and Director of ProPEL (Professional Practice, Education and Learning School of Education) at the University of Stirling, UK. Her research has focused on lifelong learning and education in the everyday activity of 'workplaces' and organizations, with particular interest in understanding how identities, power relations and knowledge emerge in the rapidly changing conditions of globalized workplace practices. Her remit at Stirling is broad and interdisciplinary: to promote innovative studies of professional knowledge, practices and learning across domains such as health care, management, social services and education, and to explore effective new approaches to support professional learning across contexts of higher education, workplace and community.

Leif Jarle Gressgård is a senior researcher at the International Research Institute of Stavanger (IRIS) in Bergen, Norway, and holds a PhD in Information Management from the Norwegian School of Economics. His primary research interests are in the fields of automation systems and decision support, computer-mediated communication, innovation and creativity management and organizational learning and safety.

Kåre Hansen is Research Director at International Research Institute of Stavanger (IRIS) in Bergen, Norway. He has conducted several projects focusing on various aspects of working life in Norway, especially concerning employee participation, working environment and cooperation between unions and management. He is currently managing a project studying conflict management in Norwegian knowledge companies.

Cathrine Hasse is Professor in the Department of Education, Aarhus University, Denmark. She holds a PhD in Anthropology from the University of Copenhagen and has studied organizational cultures in academia for more than 15 years. She has worked on how culture influences learning at physics institutes in five European countries, and in the project 'The Cultural Dimensions of Science' she compared physicists learning, teaching and researching in Denmark and Italy. She takes a special interest in how technologies become entangled with cultural learning processes and how cultural models both include and exclude people and artefacts from organizations.

Steen Høyrup is Associate Professor and Director of the research program Organization and Learning in the Department of Education, Aarhus University, Denmark. His research interests include employee-driven innovation, organizational learning and change, competence development and reflection and learning.

Natasha Kersh is Research Officer and MA tutor at the Faculty of Policy and Society, Institute of Education, University of London. Her research has focused on workplace learning, employability and post-compulsory education. She undertook her PhD research in Comparative and International Education at Oxford University, and is now a member of the Economic and Social Research Council Centre for Learning and Life Chances in Knowledge Economies and Societies (LLAKES). She is researching aspects of adult literacy and workplace learning.

Peer Hull Kristensen is Professor of the Sociology of Firms and Work Organization in the Department of Business and Politics, Copenhagen Business School. His research interest is the comparative study of national business systems, labour markets, the organization of multinational companies and the ongoing mutations of capitalisms. His current focus is on how changing forms of work organization enable new strategies for firms globally, and how this in turn is made possible by making novel use of institutions and creating novel institutional complementarities.

Tea Lempiälä is a researcher at Aalto University, Helsinki, Finland. She holds a DrSc (Economics) from Aalto University, School of Economics. Her main research interests include practices and processes of innovation, collaborative idea development, interdisciplinary research and sustainability.

Maja Lotz is Assistant Professor in the Department of Business and Politics, Copenhagen Business School. She is a post-doctoral fellow at Stanford University. Rooted within the field of Organizational Sociology, she researches into new forms of work organization, co-creation, learning and experimental governance within and across various organizational settings (such as teams, units and firms). In particular, she has studied the dynamics of everyday work roles, communities and work-organizing practices enabling mutual learning and innovative co-creation in current economic organizations.

Kirsten Møller holds a Master's in Psychology from Copenhagen University, Denmark. She is Coordinator of the European Research Network, EDI-Europe. This network has its secretariat at Aarhus University, Denmark. She has worked as an advisor on labour market and educational issues in the Ministry of Employment in Denmark and in the Danish Trade Union Movement. Earlier she worked in Nordic and international teams of researchers at the Nordic Council of Ministers and at the former Danish University: The Royal Danish School of Educational Studies. Her research interests have included socio-psychological stress, gender related workplace behaviour, employee-driven innovation, social innovation and organizational learning and change.

Oriana Price has worked in the private, public and higher education sectors for the past 20 years. In these sectors she has held numerous management and specialist positions and led a number of organizational change initiatives. Having completed her BA in Psychology and her MBA, Oriana is now completing her doctoral studies in the area of organizational change and practice at the University of Technology, Sydney.

Renaud Redien-Collot is Director of International Affairs at Novancia, Paris. His research interests include minority entrepreneurship, theories of innovation, leadership and intrapreneurship, entrepreneurship education, epistemology of the praxis in the realm of entrepreneurship and the development of public and private discourse about entrepreneurs and entrepreneurship.

Hermine Scheeres is Adjunct Professor in the Faculty of Arts and Social Sciences, University of Technology, Sydney. Her research focuses on communication and culture, organizational learning and organizational change. Her publications span the fields of adult education language and discourse, and work and organizational studies. Recent research projects include the investigation of learning and change in public and private organizations, and communication in hospital emergency departments.

Amir H. Taherizadeh is currently a research fellow at Georg Simon Ohm University of Applied Sciences, Nuremberg (Germany). Amir holds an MBA in International Business and Innovation Management from the University of Malaya (UM) and a Master's in Language Education (TEFL) from the

University of Isfahan. Amir's main research interests include innovation management, internationalization of SMEs, strategy and organization.

Ann-Charlotte Teglborg is Professor and Researcher at Novancia, a business school of Paris dedicated to entrepreneurship and business development. Ann holds a PhD in Management from Paris I Sorbonne-IAE de Paris and HEC Paris. Her main research interests include employee-driven innovation and open innovation.

Guy Van Gyes obtained degrees in Political Sciences (1992) and History (1991) at the Catholic University of Leuven (K.U. Leuven). He started his research career in the Department of Sociology at K.U. Leuven. In 1993, he became a research associate at HIVA. Since 1999, he has been a research manager at HIVA in the work and organization sector. His research field includes industrial relations, employee participation and organizational development.

Céline Viala is a researcher at Arcos, France. Céline holds a PhD from Paris-Dauphine University. Her main research interests include corporate entrepreneurship and innovation management. She has specialized in qualitative methods.

Edmund Waite is a researcher at LLAKES (Centre for Learning and Life Chances in Knowledge Economies and Societies), Institute of Education, London University. He holds a PhD in Social Anthropology from Cambridge University. His research interests and publications relate to the study of adult literacy in the UK and workplace learning, as well as the anthropological study of education in Muslim societies.

Sari Yli-Kauhaluoma is a research fellow in Organization and Management at the Aalto University, School of Economics in Helsinki, Finland. Her main research interests include collaboration dynamics and technology-related practices in professional communities. Her work has appeared in *Organization*, *International Small Business Journal* and *Time & Society*.

Abbreviations

ABSWL	Adult Basic Skills and Workplace Learning
ANT	Actor Network Theory
Ba	a group context where knowledge is shared, generated and put into practice through collaboration
CEO	chief executive officer, managing director, the highest-ranking corporate officer
CERI/STI	Centre for Educational Research and Innovation/Science, Technology and Industry
CESifo	Centre for Economic Studies
DFEE	Department for Education and Employment
DUI-mode	Doing Using Interacting mode
EDI	employee-driven innovation
EDUADM	administration of educational affairs
EO	entrepreneurial orientation
ESOL	English for speakers of other languages
E-tools	electronic tools
EU	European Union
EWC	European Works Council
EWCS	European Working Conditions Survey
GCSE	General Certificate of Secondary Education
HET	Suzhou Dushu Lake Higher Education Town
HR	human resources
IALS	International Adult Literacy Survey
ICMI	Innovation Capability Measurement Instrument
ICT	information communication technology
IDP	integrated development practices
IfM	Institut für Mittelstandsforschung (Bonn)
IR	industrial relations
IWB	innovative work behaviour
LAICS	leadership and innovation in complex systems
LLAKES	Centre for Learning and Life Chances in Knowledge Economies and Societies, University of London
LO	Landsorganisationen i Danmark (The Danish Confederation of Trade Unions)
LSM	learning support managers
MBA	Master of Business Administration
MNC	multinational corporations
NACE	classification of economic activities in the European community

NS	not significant
OECD	Organisation for Economic Co-operation and Development
PC	personal computer
PVC	poly vinyl chloride
PVDF	a special kind of monomer renowned for its technical characteristics
QIA	Quality Improvement Agency
QPS	quality production systems
R&D	research and development
RBV	resource-based view
S&T	science and technology
SFL	skills for life
SIP	Suzhou Industrial Park
SME	small and medium-sized enterprises
SOADM	administration of social affairs
STS	Southern Transport Systems
UK	United Kingdom
ULR	union learning representatives
UNU-MERIT	United Nations University is a research and training centre of United Nations University and works in close collaboration with the University of Maastricht
VIS	visitation office

Part I
The Nature of Employee-Driven Innovation

1
Employee-Driven Innovation: A New Phenomenon, Concept and Mode of Innovation

Steen Høyrup

The overall purpose of this chapter is to define the phenomenon *employee-driven innovation* and to unveil the nature of these innovation processes in the context of the workplace. This venture involves an investigation into the relations that link the worker, everyday practices in the workplace, workplace learning and employee-driven innovation. It is especially important to clarify the question: How do employees who are learning in the workplace produce innovation? Selected learning theories – linking workplace learning and employee-driven innovation – are presented and discussed to illuminate this question. The theories include both individual learning and organizational learning, and include classic theory as well as quite new theories presented in this volume.

The fundamental approach in the learning theories is that employee learning at the workplace – in terms of new knowledge, expertise and problem-solving skills – constitutes the raw material for employee-driven innovation. Basically, both employee initiatives and autonomy, on the one side, and the structure and conditions of work, on the other side, are important for innovation. What is the workplace like that supports learning and innovation of employees?

Although the concept of employee-driven innovation is quite new, some antecedents of the concept are identified. A comprehensive model that conceptualizes the phenomenon in a broader organizational and global context is presented. This model may serve as a framework including issues, perspectives and level of analysis applied and discussed in the present volume.

To illuminate the relation between work organization and innovation and learning, international comparative and quantitative research is consulted. The perspective here is the macro-level.

At the micro-level, the application of the learning theories seems to indicate that learning and innovation processes are closely interwoven and to a high degree possess the same characteristics. The theories are applied to shed light on the question: What are the organizational conditions that facilitate learning and innovation in organizations? This issue forms the departure point for conceptualizing *divergent and convergent processes* – as conceptual tools for analysing how innovation and learning are related within an organizational context.

It is concluded that the processes of learning and innovation are closely interwoven and constitute preconditions of each other. Thus, it is practically impossible to separate innovation and learning processes. Finally, it is concluded that a very important relation in our perspective of employee-driven innovation is that *learning can produce innovation*. In this production of innovation, adaptive and innovative learning are prominent processes among a more comprehensive range of learning processes – supported by convergent and divergent processes in the organization – that contribute to and produce innovation. This is realized in a complex interplay of processes that include factors at the individual level as well as organizational culture.

Innovation: what is new?

For more than half a century, innovation and innovation processes have been addressed by dedicated researchers, politicians and practitioners trying to unveil and apply the secrets and potential of innovation, and trying to develop new research-based knowledge in the field. Thus, the phenomenon of innovation is not itself new. But it is new that innovation today is an imperative for societies and workplaces: *workplaces have to be innovative*. As Fees and Taherizadeh state:

> In a borderless world where boundaries are no longer relevant, where accessibility to knowledge, expertise and technology is far advanced; innovation has become a necessity rather than a luxury (this volume, p. 188).
>
> And further: machines do not innovate, but human beings do! It may, therefore, be the case that at the present turbulent time, when firms are pushed to innovate despite being bounded by their limited resources and capabilities, they need to turn their focus within and utilize the resources they already possess... their invisible assets or employees. And – of great importance in the perspective of the global scene: Empowered and qualified employees are one of those intangible and dynamic capabilities which cannot be imitated by competitors easily. (Unpublished paper)

It becomes obvious that the importance of employee-driven innovation in society and workplaces today is firmly rooted at the societal level, the organizational level and the individual level as well. The concept of employee-driven innovation may thus be connected to concepts such as knowledge

society, network society and human capital, and international political discourses in, for example, OECD.
Important political discourses are presented in the EU and OECD. In an OECD report from 2010, the following statement is emphasized:

> Innovation is widely recognized as an important engine of growth. The underlying approach to innovation has been changing, shifting away from models largely focused on Research and Development (R&D) in knowledge-based globalised economies and giving more emphasis to other major sources of the innovation process. Understanding how organizations build up resources for innovation has thus become a crucial challenge to find new ways of supporting innovation in all areas of activity. (CERI/STI, 2010)

That states exactly what this volume is about: to emphasize and investigate employees as being the most important factor in the before-mentioned category of 'other major sources of the innovation process', and to investigate how organizations build up resources for innovation

In this context, we find a new trend in the development of the very concept of innovation, moving from a focus on new products related to economic benefit (Schumpeter, 1934) – especially in the field of high technology – to conceptualizations underlining current configurations of daily work practices and the current search for new resources for innovation, in both the private and the public sector. In this perspective, innovation – as a broad range of processes and products – is not conceptualized as separate units, but as embedded in daily work activities and job enactment and social processes in the organization. As Price *et al.* state:

> We take the view here that innovation arises from the everyday cultural practices of workers – the ways workers enact their jobs, interact with each other and seek to become fuller members of their organizations. It occurs through workers finding ways of meeting their own interests and desires as well as those of their employers. (This volume, pp. 77–78)

This approach underlines the organization–individual interaction, and here innovation arises from the everyday cultural practices of workers. It is, of course, important to underline characteristics of the organization and its development as an important context for innovation processes. This context may foster or impede innovative processes.

An important trend in organization development and structure relevant to this issue is that organizations in a global competence context may choose to decentralize responsibilities and delegate planning and execution to operatives, often working in teams. The aim is to create a modular form of organization that can easily be changed and in which the responsibility for continuous improvement can be handed over to operative teams and cross-sectional ad hoc teams. In this process, even external relations among firms and institutions may become decentralized to employees, implying that the traditional hierarchical authority becomes undermined. But these changes run counter to a multiplicity of vested interests, to a number of institutionalized routines, and are dependent on whether employees are in fact able and ready to take on new responsibilities. An important factor here is a continuous upgrading of skills through both formal and informal learning in the workplace – that is, workplace learning that supports the employees' innovative potential.

From this perspective, it is obvious that the relationship between workers and organizations is crucial. Price *et al.* mention (this volume, p. 77) that this relationship – and the nature of organizations – has changed dramatically. The changes in organizations include flatter structures, greater autonomy, less formalization, and joint ventures to serve global markets. Boud states that contemporary workers are not only expected to take responsibility for their own learning, development and careers, but they can also expect to take responsibility and ownership for work activities not defined by their organizations or bound to the job they were employed to do. Employees do not necessarily have to wait for managers to take the initiative. This opens the way to a creative involvement in the organization that can position workers as drivers of innovation (Price *et al.*, this volume, p. 77). These new characteristics of organizations and the organization–employee relationship also open up possibilities for workers to be self-governing and creative, both of which are important prerequisites for workplace learning.

The new development trends that we observe in society, in the organization of companies and in the relationship between company and employee – combined with the new ideas of innovation – contribute to creating the concept of *employee-driven innovation*. The recognition of ordinary employees as being important and continuous producers of new knowledge, new ideas and solutions, and who are continuously remaking their jobs, comprises a new and fruitful perspective on innovation in organizations.

Finally, it should be mentioned that a learning perspective on innovation – especially employee-driven innovation in a learning perspective – comprises a new approach to innovation. In the new conceptualizations of innovation and resources of innovation – ordinary employees – the learning perspective seems to offer a promising basis for knowledge production in the field of innovation.

Based on the development of organization–employee relations as outlined above, the obvious approach is to look for new resources and drivers of innovation in order to tap the full potential of innovation, and to add, support and realize new – or maybe hidden – resources of innovation. This is the perspective from which employee-driven innovation has to be advanced.

Ordinary employees of the organization constitute a strong, but often overlooked, resource of innovation.

How to enhance innovative working, and innovation and learning at work, thus continues to be one of the most significant challenges for organizations. These are the challenges that are going to be analysed in the present chapter.

Conceptualizing employee-driven innovation

Employee-driven innovation is an umbrella concept that covers a broad range of innovation processes and issues. It is in keeping with the 'classic' innovation concept – in several dimensions. Employee-driven innovation refers to both process and product, and, as with any other kind of innovation, the result may be substantive products, services and/or processes of an organization, and generation and/or implementation may be involved in the process. What is new compared with more classic innovation conceptualizations is that employee-driven innovation may further contribute to innovation in ways that are subtle, informal, in ways that are not a part of the organization's explicit goals and strategies, and are, finally, beyond the reach of managers. This form of employee-driven innovation consists of practices that may not be initiated with a goal of innovation in mind, but nonetheless have it as a central outcome (Price *et al.*, this volume, p. 78).

How, then, should employee-driven innovation be defined?

Gyes *et al.* propose this definition:

> Employee-driven innovation is a form of direct participation in which the employee takes the initiative to develop, propose and implement change. (This volume, p. 232)

LO, The Danish Confederation of Trade Unions, proposes:

> Employee-driven innovation means that employees systematically and actively contribute to the generation of new ideas which create value when they are implemented. (LO, 2008: 10)

Kesting suggests this conceptualization:

> EDI (Employee-driven Innovation) here refers to the generation and implementation of new ideas, products, and processes originated by a single employee or by joint efforts of two or more employees... Therefore EDI indicates that innovation emerges coincidentally among 'ordinary employees' from shop floor workers, professional, and middle managers across the boundaries of existing departments and professions. (Kesting and Ulhøi, 2008: 2)

A trait common to the first-mentioned definitions is that the notion is conceptualized by relating it to a broad social process: *direct participation*. The weakness of this approach is that it is a precondition that this concept has a shared and precise meaning, and this may not always be the case. Kesting's definition includes important characteristics of the innovation concept: the content and products of the innovation may be ideas, products and processes. Finally, we get a conceptualization of what is called 'ordinary employees' or workers: these are members of the organization who are not formally assigned the task of being innovative. Kesting uses the term that the new ideas, and so on, are 'originating from' employees. This seems to be a good expression for many reasons. It covers a broad range of creation and utilization of ideas and changes of work practices. The creation could be accidental; it can be informal, that is, without intent; it may take the shape of contributions in an active and systematic way (as in the LO definition); and it may be the active use of management of employees' ideas and knowledge, when management *invites* employees *to participate* in innovative processes. We would therefore prefer Kesting's definition to be slightly extended:

> Employee-driven innovation refers to the generation and implementation of new ideas, products, and processes – including the everyday remaking of jobs and organizational practices – originating from interaction of employees, who are not assigned to this task. The processes are unfolded in an organization and may be integrated in cooperative and managerial efforts of the organization. Employees are active and may initiate, support or even drive/lead the processes.

In this conceptualization, employee-driven innovation indicates a *bottom-up process* in the organization. *Employee-driven innovation begins at the job and at worker level*. But the situation is rather complex: innovation – including employee-driven innovation – is basically a *social process*. As Sundbo states: 'single individuals alone do not create innovation within modern firms' (Sundbo, 2003: 101). Employee-driven innovation may begin at the job and worker level, but turns out to be an interaction between several actors that may include employees and managers as well. Therefore it may be difficult to isolate employee-driven innovation as a pure bottom-up process in the organization, and the processes may not be isolated from management, but related to management. In some conceptualizations of employee-driven innovation, the management perspective seems to be dominant (Kesting and Ulhøi, 2010). Management plays a role in employee-driven innovation, in several ways. Sundbo states that top management can rarely get the innovative ideas they need themselves and they cannot just rely on scientific or expert lines of thinking

(Sundbo, 2003: 104). This is quite in accordance with Kesting and Ulhøi, who state:

> Employees typically acquire exclusive and in-depth and highly context-dependent knowledge that managers often do not possess. (Kesting and Ulhøi, 2010: 71)

This is an important argument for the importance and indispensability of employee-driven innovation in modern organizations.

It may be possible for the employees in units of the organization – for instance, in teams – without interference from management to continuously develop and remake their daily work practice and to conduct other kinds of innovations. But that also implies a threat: if the organization is overflowing with uncoordinated ideas and solutions from employees, this may threaten to tear apart the very order of the system, and the process of change from this perspective may therefore appear anarchic and unpredictable. Consequently, managers may be involved in employee-driven innovation on two levels. First, they may join the process of employee-driven innovation to coordinate and systematize the processes initiated by the employees, to use the resources of the employees and make these processes constructive to the organization. Cooperation between management and employees is thus essential (LO, 2008: 9). Employee-driven innovation in this case is a mixture of bottom-up and top-down processes in the organization (Teglborg, 2010a).

Second, managers may take the initiative in innovation by *inviting employees to participate* in innovative processes, and the involvement of employees in the innovation process may be formally organized. This is clearly a top-down process. This understanding of the concept of employee-driven innovation is prominent in approaches centred on the term participation, as in the approach by Kesting and Ulhøi (2010). They state that 'there is one last management task that has resisted increased participation: decisions about major innovations [p. 66]. Employee-driven innovation is apprehended as employee participation/involvement in innovation-related activities and decisions [p. 67].'

It may be questioned whether this form of top-down innovation should be included in or excluded from the concept of employee-driven innovation. In the present book, we shall prefer to conceptualize employee-driven innovation both in a narrow sense – referring to bottom-up processes and the mixture of bottom-up and top-down processes – and in a broader sense that includes the top-down process as well.

Thus, it appears fruitful to distinguish between:

First order EDI (Employee-driven innovation): These are bottom-up processes. Innovation arises from the everyday cultural practice of workers – the ways workers enact their jobs, interact with each other and seek to become fuller members of their organizations (Price *et al.*, this volume, p. 77). It is a reconstruction of work practices that is not initiated with a goal of innovation in mind, but which results in innovation as a central outcome (Price *et al.*, this volume, p. 77). Price *et al.* state that these innovations may be subtle changes that may not appear under the banner of organizational innovation, but nevertheless the changes may be substantial (Price *et al.*, this volume).

Second order EDI: a mixture of bottom-up and top-down processes in which management strives to systematize and formalize significant innovation processes initiated by employees, for example those mentioned above.

Third order EDI: a top-down process in which managers invite employees, and involve employees in participating in innovative processes, for example in terms of involvement in developmental projects, project groups in the organization, and so on.

It becomes clear that the concept of employee-driven innovation covers a broad range of complex processes that can be related to both a formal–informal dimension and a bottom-up/top-down point of view. This book covers all aspects of this complex phenomenon. Furthermore, employee-driven innovation may be conceptualized as embedded both within working processes – the daily work practice – and within learning processes, especially workplace learning.

Historical antecedents of employee-driven innovation

Referring to Alter, Teglborg states that the industrial history of the nineteenth century is full of examples of sometimes decisive innovations developed by ordinary workers (Alter, 2000; Teglborg, 2010b).

From a theoretical perspective, we find evidence of employee-driven innovation back in the writings of Peter F. Drucker in 1987, reflecting on *Social Innovation – Management's New Dimension* (Drucker, 1987). He states that the essence of modern organization is 'to make individual strengths and knowledge productive and to make individual weaknesses irrelevant' (ibid.: 33). Compared with what is called traditional organizations, these organizations were characterized by the following: everybody did exactly the same unskilled jobs in which sheer strength was the main contribution. *The knowledge that did exist in the organization was concentrated at the top of the organization and in very few heads.* By comparison, in modern organizations everybody has fairly advanced knowledge and skills. All are contributing their knowledge. 'The little each knows matters' (ibid.: 33). In this early text, Drucker underlines the diversity of employees, that knowledge is not concentrated at the top of the organization, that all are contributing their knowledge, and what each employee knows matters. This conceptualization of organizations can be seen as an antecedent of the concept of employee-driven innovation. But Drucker's perspective is that of organization and management. We have to add and underline the perspective of the employees in our understanding of the phenomenon.

In the book *How Breakthroughs Happen. The Surprising Truth About How Companies Innovate*, the core of Hargadon's thinking fits very well with the essence of the concept of employee-driven innovation and provides inspiration for this concept (Hargadon, 2003). Two concepts are basic. First, the concept of *bridging* implies that 'break-through innovation depends on exploiting the *past*' (Hargadon, 2003: VII). In the case of employee-driven innovation, this may be the experiences and knowledge of the employees. 'Innovation is the result of synthesizing, or "bridging" ideas from different domains' (ibid.: VIII). 'Combining often well-known insights from diverse settings creates novel ideas that can, in turn, evolve into innovations' (Hargadon, 2003: VII). In our case, this may be combining ideas and expertise of employees.

Kathleen Eisenhard, in the Foreword of the book supplement, provides a statement of great relevance for employee-driven innovation:

> Moreover, innovations that rely on the past are pragmatic. They save developers and their manager's time and money, even as they lower risk. In contrast, innovative attempts that focus on developing fundamentally new visions from entirely novel knowledge very often fail. This path is too slow, too challenging, and too risky. (Hargadon, 2003: VII)

Hargadon himself states it even more succinctly:

> The result is an innovation process that thrives by making smaller bets, by building the future from what's already at hand. (Hargadon, 2003: XII)

These citations underline the huge potential and strength of employee-driven innovation.

The second basic concept in Hargadon's thinking is 'building communities'. Quite in accordance with the concept of employee-driven innovation, Kathleen Eisenhard states: 'Breaking through innovations depends on ordinary people, bridging their expertise and building communities around their insights' (Hargadon, 2003: X). Basic is the concept of recombination, rather than invention.

The concept of breaking through innovations is thus very much in accordance with the basic philosophy behind employee-driven innovation.

Central to the very concept of employee-driven innovation is our basic understanding of work, learning and innovation as basic human activities. Working, learning and innovating are basic human activities and can be

12 Employee-Driven Innovation

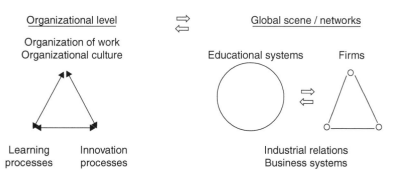

Figure 1.1 EDI in an organizational and global context

seen – as Brown and Duguid state it – as closely related forms of human activities, and potentially complementary, not conflicting, forces (Brown and Duguid, 1991: 40). In this view, learning is not separated from working, and learners are not separated from workers. Learning is understood as the bridge between working and innovating (ibid.: 41).

How can the new phenomenon of employee-driven innovation be characterized? The effort here is to unveil core characteristics of the phenomenon, conceptualize underlying processes and comparable concepts, and locate the concept in a more comprehensive context. A basic figure is presented above, showing the core elements of employee-driven innovation elaborated in the chapters of this book: interaction between innovation and learning processes, in the context of the organization in the global scene. This approach includes the application of different perspectives to such basic concepts as working, learning and innovation.

On the left side of the figure, we find the single workplace conceptualized in organizational terms. Important interactions and relations here involve the three elements: organization of work/organizational culture, learning processes and innovation processes. The workplace/organization is a player in the global scene in which firms are connected in networks and interact with educational systems (right side of the figure).

Connecting work organization to learning and innovation: macro-level research

A CERI/OECD report from April 2010 (CERI/STI, 2010) presents an interesting survey of the literature on learning organizations. It states that there is nothing like a unified definition or concept of the learning organization, but some common definitional ground has been identified. 'Learning organizations are those with a capacity to adapt and compete through learning' (ibid.: 3). The multi-level nature of the concept and the role of learning cultures – understood as beliefs, attitudes and values supportive of employee learning – are emphasized. Learning organizations are organizations in which high levels of autonomy in work are combined with high levels of learning, problem-solving and task complexity (ibid.: 4). The learning organization is important in this context because the notion of the learning organization is closely linked to that of organizational learning, and scholars such as Senge (1990) and Argyris and Schön (1978) have long maintained that organizational learning promotes creativity and innovation (CERI/STI, 2010: 7–9). Below we refer to a thorough and long-lasting piece of research by Lorenz and Valeyre (2005) which presents empirical evidence on the spread and characteristics of learning organizations at national and EU levels. An important question to be clarified is: What kinds of work organization support a culture of innovative behaviour, creativity and learning?

The research by Lorenz and Valeyre is quantitative and comparative, focusing on work organization and employee learning dynamics in private sector establishments of EU member states (CERI/STI, 2010: 7–9). The research draws on the 2000 and 2005 waves of the European Working Conditions Survey (EWCS) carried out at the individual level by the European Foundation for the Improvement of Living and Working Conditions for the EU15 and EU27, respectively (CERI/STI, 2010: 14).

Some results of this research should be presented briefly here, based on the CERI/STI report. Lorenz and Valeyre build on concepts of four basic systems of work organization that support different styles of leaning and innovation:

Operating adhocracy, also called *Discretionary learning form of work*, relies on the expertise of individual professionals and uses project structures to temporarily fuse the knowledge of experts into creative project teams that carry out innovative projects, typically on behalf of its clients. High levels of discretion in work provide scope for exploring new knowledge, and adhocracies tend to show a superior capacity for radical innovation (CERI/STI, 2010: 14).

The lean model of production is a relatively bureaucratic form that relies on formal team structures and rules of job rotation to embed knowledge within collective organization. Stable job careers within internal labour markets provide incentives for members to commit themselves to the goals of continuous product and process improvement, and the J-form tends to excel at incremental learning and innovation (CERI/STI, 2010: 15).

Taylorism, named for the industrial engineer F. W. Taylor (1856–1915), includes principles of organization and production efficiency, notably breaking actions and tasks into small and simple segments (maximum job fragmentation), excessive external control of work processes, and separation of work execution from work planning.

The traditional organizational form, based on simple management structure.

Empirical results are:

Discretionary learning form of work. This organizational form is distinctive for the way high levels of autonomy in work are combined with high levels of learning, problem-solving and task complexity (CERI/STI, 2010: 16). Further, monotony of tasks and repetitiveness of tasks are very low.

Lean production form is – compared with the above organizational form – characterized by low levels of employee discretion in setting work pace and methods. The use of job rotation and teamwork is very high. Learning – in terms of learning new things at work – is ranked relatively high, compared with Taylorism and the traditional organizational form, but a little below the discretionary learning form of work, which has the highest ranking. Problem-solving activities are ranked very high.

Taylorism. In most respects, the work situation is the opposite of that found in *Discretionary learning form of work*, with low discretion and low levels of learning and problem-solving.

Traditional organization. We find the lowest levels of learning and complexity of tasks in this organizational form. Monotony of work is close to average.

It seems obvious from the empirical data that the *Discretionary learning form of work* fosters learning at work most successfully, followed by the lean form. Taylorism and traditional organization rank lowest in this respect.

The linking of organizational form to learning is interesting, but empirical data also reveal where – in nations and within industries – the different organizational forms are prevalent.

The discretionary learning form of work organization is especially prevalent in several service sectors. The lean model of production is more developed in the manufacturing sector. The Taylorist form is notably present in textiles, clothing and leather products, food processing, wood and paper products and transport equipment. The traditional organizational form is found principally in the services, notably land transport, personal services, hotels and restaurants, post and telecommunication, and wholesale and retail trade (CERI/STI, 2010: 16).

Empirical data show that there are wide differences in the employee learning dynamics across European nations. The discretionary learning forms of work organization are most widely diffused in the Netherlands, the Nordic countries and to a lesser extent Germany and Austria, whereas they are least diffused in Ireland and the southern European nations (CERI/STI, 2010: 17). The more bureaucratic lean model is most prevalent in the United Kingdom, Ireland and Spain and to a lesser extent in France (CERI/STI, 2010: 16). The low-learning Taylorist forms of work organization show almost the reverse pattern compared with the discretionary learning forms, being most frequent in the southern European nations and in Ireland and Italy. The traditional forms of work organization are most prevalent in Greece and Italy.

It can be concluded that the Nordic nations and the Netherlands stand out for their frequent use of the 'discretionary learning' forms of work organization and their low level of use of Taylorism (CERI/STI, 2010: 20). It should also be underlined that discretionary learning shows the highest score in the categories 'Almost always or often applies one's own ideas in work' and 'Strongly agrees or agrees that the employee has opportunities to learn and grow at work' (CERI/STI, 2010: 21).

This leads us to the next question: what are the learning processes like that constitute discretionary learning?

Learning theories: linking learning and innovation

It seems that 'learning' often appears as a black box: the micro-processes in 'discretionary learning', as mentioned above, are not made explicit. And literature on learning and innovation often refers to the concept of *DUI-mode*, coined by Jensen *et al.* (2007): an experience-based mode of learning and innovation, based on doing, using and interacting. It relies on informal processes of learning and experience-based know-how (Høyrup, 2010). But what is the nature of this interaction of learning and innovation processes? That is the central issue in this section.

Initially a working definition of workplace learning should be stated, as a shared frame of reference. Subsequently, the authors in the separate chapters are going to elaborate further on these key concepts.

Referring to Karen Evans, learning may be conceived as *'the processes by which human capacities are expanded*, and we understand "workplace learning" as processes by which human capacities are expanded in, for and through the workplace' (Evans *et al.*, 2006).

The 'expansion of human capacities' may be understood in terms of knowledge and behaviour; referring to this, learning may be defined as 'a systematic change in behaviour or knowledge informed by experience' (Miner *et al.*, 2001: 305). These definitions of learning are valid for both the individual and the organization.

In the following paragraphs, the focus is on describing and analysing the relation between innovation, learning and organization of work, referring to the figure on page 12.

Remaking everyday work practice

The views on working, learning and innovation as basic and closely related human and social processes, stated by Brown and Duguid (1991: 40), are characteristic of the two different approaches presented by the Australian researchers Stephen Billett and David Boud (this volume, p. 77–91 and 92–107).

Both researchers have worked extensively – both theoretically and empirically – within the same perspective: to conceptualize workplace learning and innovation – especially employee-driven innovation – in what is termed 'remaking everyday work practice' or 'the reconstruction of work practice' or 'remaking of cultural practices' (Billett *et al.*, 2005).

A trait common to the two approaches to innovation is that innovation processes can be current, emergent, spontaneous, informal and unplanned; they may not be part of the explicit agenda of the organization, or necessarily something which managers are deeply involved in initiating. Innovation and learning processes are interconnected and integrated processes. Innovations may be incremental but may have a huge impact on the organization as a whole.

Both authors underline workplace learning and change initiatives of employees as a basis for workplace innovation. Yet the two learning approaches applied are different.

Billett coins the term *a socio-personal account* as a way to understand employee-driven innovation. The basic dynamics of employee-driven innovation from this perspective is the *relational interdependence* between two sets of factors: (1) individual engagement in *work*, how individuals elect to participate in work. These terms denote the workers' personal *agency* in the workplace setting; (2) *workplace affordances*, how individuals are permitted to participate in work (Billett, this volume). These two aspects comprise a duality that is interdependent and relational.

Innovation from this perspective comprises the processes through which workers learn and actively remake their occupational practices. In this way, the concept of innovation connects individual learning and development of the firm. In this view, employee-driven innovation – like innovation in general – includes important personal and socially derived purposes, and the processes securing and sustaining innovation at work draw on personal and social contributions. Billett emphasizes that the key premise of duality comprising relational interdependencies is that neither the social suggestion nor individuals' agency alone is sufficient to promote learning and the remaking of the cultural practices that constitute work (Billett *et al.*, 2005).

Innovation at work is conceptualized as being comprised of two kinds of change processes that occur interdependently of each other: (1) individual learning and (2) the transformation of workplace practices. The changes occur in relation to each other. In this conceptualization, innovation is a broad concept that includes learning processes. In this view, everyday processes of thinking and acting at work are constructive acts through which work tasks and processes are reconfigured in response to new requirements arising at a particular moment in time and in response to specific situational requests or problems, such as work tasks. These everyday processes of work-related thinking and acting lead to the remaking of occupational practice and comprise what is called employee-led innovation (Billett, this volume). And, further, in confronting and responding to significant changes (innovations) of practice, both new learning for those enacting these tasks and the transformation of practice co-occur. In this way, learning is both a response to innovative processes and an inspiration for innovative processes of the firm.

Like all innovation, innovation at work comprises both *processes* and *outcomes*. One important *outcome* of innovation at work is that the viability of the enterprise is sustained in the face of changing internal and external challenges. Innovation processes comprise a process through which employees come to learn and actively remake – reconstruct – their occupational practices. We may see these reconstructions as incremental innovation processes. In this view, innovation processes and learning processes seem to be integrated, seem to be two aspects of the same social process: in the work process, employees act and learn. Learning includes a change in behaviour.

Despite the term *employee-driven innovation*, these processes should not be looked upon as isolated processes, but as phenomena in the larger workplace setting. We have to include the sometimes necessary and sometimes hindering way in which workers' personal agency is bounded by workplace settings: that is, the degree to which workers are able to exercise their agency, within the constraints provided and made permissible by the workplace (Billett, this volume). The exercise of unbounded individual agency is not possible, yet the employee may show initiatives at the workplace. In this approach, Billett underlines the important themes stated in terms with slightly different meanings: variety of work, autonomy and discretion, factors that in many theories and empirical studies are explained as crucial factors determining learning and different learning forms, such as adaptive learning and innovative learning, concepts used and elaborated by Per-Erik Ellström.

To sum up: Billett's conceptualization of employee-driven innovation underlines the perspective of the employees by using the term *individual agency*. Yet this is not an individual psychological approach, because individual agency is seen as related interdependently to the workplace affordances. This also means that, despite the term *employee-driven innovation*, this phenomenon is not seen in isolation but as part of a complex workplace system. In the thinking of Billett, innovation and learning are closely related: 'innovations are as much about human learning as about the implementation of new practices, and one without the other is unlikely' (Billett, this volume).

Enmeshing elements of existing practices

Price *et al.* also see employees as crucial to innovation at work, and underline the perspective of informal, everyday ways in which workers think about, construct and reconstruct their work (this volume, p. 77). Price *et al.* build on Fenwick's idea that innovation is a form of 'everyday practicing Learning' (Fenwick, 2003; Price *et al.*, this volume) and elaboratethis idea further, stating that it is through a 'process of enmeshment of practices that we understand how employee-driven innovation emerges' (Price *et al.*, this volume):

> Whether consciously or unconsciously, as they enact organizational practices workers draw on their practical intelligibility to carry practices forward and at the same time vary them, enmeshing elements of existing practices with previous understandings of similar practices from other contexts... (Price *et al.*, this volume, p. 81)

'Employee-led innovation occurs through a conjunction of circumstances and the ability of employees to respond to and utilize the opportunities that occur at work' (Price *et al.*, this volume). We can see the same relational structure as in Billett's thinking in the relation between engagement of employees and affordances of the work, but Price *et al.* prefer the slightly different concepts of *practical intelligibility* – borrowed from Schatzki (2002) – and *enmeshing elements of existing practices*.

In describing the remaking of jobs and innovation through the everyday work of employees, Boud calls attention to tensions and gaps between new and old ways of doing things. Empirical data show that workers went about creating their jobs by bringing together existing and new organizational practices. 'The context of their organizations was such that workers were afforded the freedom (and perhaps expected) to become the architects in building bridges to fill these gaps, in finding ways to smooth the tensions...' (Price *et al.*, this volume, p. 88).

Learning is a necessary and important feature in this kind of innovation. Price *et al.* (this volume) address the basic complexity of the relationship between work, learning and innovation.

In this approach learning is not understood as being determined by a dichotomy between individual and social factors, but learners take up opportunities to learn that 'result from a complex combination of situational factors that generate invitational patterns signaled from and by various understandings and interactions among actors doing collective work' (ibid.: 359).

Applying the basic concepts of complexity theory in opposition to thinking that explains learning and innovation in deterministic or reductionist terms, Boud proposes the following definition of learning:

> Learning is implicated in re-making practices, and learning develops as a collective generative endeavor from changing patterns of interactional understandings with others. (Price *et al.*, this volume, p.)

By this approach, Price *et al.* underline learning and innovation processes as being emergent and spontaneous, implying that we have to do with a phenomenon that cannot be completely under the control of management and organizationally planned strategies. This approach points to new concepts of management, in which management is less controlling and more participating.

In terms of spontaneous and emerging processes of learning, Price *et al.*'s concept of 'in-between' as a learning arena appears fruitful: a focus on situations or spaces where 'social' and 'work' overlap in terms of space and time, such as breaks in tea rooms. These spaces may be understood as 'hybrid spaces' (Solomon *et al.*, 2006: 5): 'that is, at one and the same time, work and socializing spaces where the participants are both working and not working'. These kinds of space provide room for new ways of being, working and learning (Solomon *et al.*, 2006: 6). For these spaces, it is important that they are neither neutral spaces nor 'power-free'; they also have to be seen in a context that includes networks of power, culture and so forth.

Boud's thoughts remind us about the complexity and changing nature of the organizational context of learning and innovation, and extend our understanding of spaces for learning and innovation: not only is formal work organization changing today, and taking a wide range of forms, but the term 'in-between' extends our view to life 'behind the scenes', to activities in more informal networks in the organization.

In the two theories presented above, innovation processes and learning processes are interwoven processes that cannot be separated and isolated in a meaningful way. This unit of processes contributes to the remaking of work practice, often in spontaneous, emergent and informal ways.

The learning form presented below – *improvisational learning* – shares some of these characteristics (the elements of non-planning and spontaneity), but it also differs: learning and innovation are related in a different way, and the learning outcome is not restricted to remaking of the job, but also includes a broader range of innovation products.

Improvisational learning: problem-solving, learning and innovation behind the scenes

Many researchers call our attention to important events and activities going on in informal networks of the organization, when we leave the area of routines and face unexpected situations and problems. This phenomenon appears to be of great importance to our field of learning and innovation. Brown and Duguid (1991: 31) and Miner *et al.* (2001) state that 'there is typically a considerable creativity and an ability to improvise when it comes to finding solutions to unexpected problems that arise.'

Sundbo states that the part of the organization which carries out the innovation activities can be conceptualized as a dual organization (Sundbo,

2003: 104). One level consists of top management, which concentrates on the general lines of the firm's development. The other level consists of employees and middle managers. This level is of special interest in relation to employee-driven innovation. This level consists of a formal organization but also develops an informal organization, a network structure that is not found in the official charts depicting the organization (Sundbo, 2003: 104). This social structure is a loosely coupled network structure, which is informal and interactive. Sundbo states:

> The network structure is a 'greenhouse' for innovative ideas and informal reflections because it constitutes a collective, a creative milieu. (Sundbo, 2003: 104)

According to Miner, improvised activities often occur outside organized routines or formal plans, and thus improvisation as a learning form is a fruitful point of departure to investigate this greenhouse and informal networks for innovative ideas and learning potential.

For Miner *et al.*, learning is experience-induced changed behaviour and/or knowledge. The authors define *learning* as a *systematic change in behaviour or knowledge informed by experience*. This experience-induced changed behaviour and/or knowledge may be retained to some degree to constitute learning.

This conceptualization relates learning to improvisation:

> We accepted the notion that pre-existing routines do not constitute improvisation; there must be some degree of novelty in the design. We took 'composition' or 'design' to mean that improvisation refers to deliberate, as opposed to accidental, creation of novel activity. (Miner *et al.*, 2001: 305)

Improvising links learning and innovation because it is both a special kind of learning and a special kind of innovation. Improvisation is defined as the deliberate and substantive fusion of the design and execution of a novel production (ibid.: 314). Improvising is a kind of learning, because improvising may involve adaptation to external events. And improvising may involve the creation of new opportunities, that is, innovative learning, and in this case creation of new opportunities is part of an innovative process as well.

Improvising is learning because it informs action:

> In improvisation, real-time experience informs novel action at the same time that the action is being taken (during the act).... The impact of real-time experience on action is the defining characteristic of improvisation. (Ibid: 316)

As mentioned above, improvisation is related to innovation, but innovation is a broader term, obviously including more than improvising. Innovation may be created through improvisation, but innovation can also be created by controlling and planning. Innovation is a necessary feature of improvisation, but this does not imply that all innovation is improvisation – *improvisation is a special type of innovation*.

In the thinking of Miner et al., improvisation can fruitfully be seen as a special type of short-term, real-time learning. Related to the definition of learning, in improvisational learning experience and related change occur at the same time. Improvisation is not just individual learning; empirical studies document that improvisation can influence long-term organizational learning, and improvisation not only draws on prior learning but may be both a special type of short-term learning and a factor that influences other, longer-term organizational learning activities (Miner et al., 2001: 306).

What then are the products – in terms of learning and innovation – of improvisation? The authors identified a broad range of improvisational productions organized into three categories:

(1) *New behaviours* (e.g. a new activity during completion of a task, or new product development processes)
(2) *Physical structures* (artefactual productions)
(3) *New interpretive frameworks* improvised by the organization. An example of the last product would be that a team reframed the meaning of the unexpected events in a novel way, and infused the prior events with new meaning (ibid.: 313).

These different products may be conceived as learning products and innovation products, and it is obvious that they may include not only the remaking of the work practice, but also a much wider range of outcomes.

What are the triggers of improvisation? Miner prefers the term improvisational 'referents':

The matter is that you can't improvise on nothing; you've gotta improvise on something. ... The referent both infuses meaning into improvisational action and provides a constraint within which the novel activity unfolds. (Ibid.: 316)

Empirical studies by Miner show that unexpected problems, temporal gaps, and unanticipated opportunities provided crucial referents that anchored and constrained the improvisational episodes (ibid.: 316). Examples show that teams often used improvisation to solve problems created by surprises in bringing a product to market (ibid.: 316). And 'in some instances, team members initially framed their activity as solving a problem, but as they improvised, they generated novel actions or interpretations that transformed the problem into a perceived opportunity' (ibid.: 316).

Improvising as learning and innovation has special qualities in terms of spontaneous, informal and emergent elements. Thus, focus on improvisational learning is on specific problems and opportunities and the specific medium or materials involved in working with them (ibid.: 316). As a kind of individual and organizational learning, improvisation can produce new cognitive knowledge for individuals and teams. In the study by Miner, teams sometimes gained new insights into technical facts or consumer preferences they had not known in advance. But the knowledge was constrained by the specific material, temporal and cognitive situation; it was situated learning, and not theoretical 'school learning'.

One important characteristic of improvisational learning – from the perspective of innovation – is that improvisational learning may have a widely varied long-term influence on organizational activities (ibid.: 319). Long-term effects may both be positive in terms of the problems solved and create valued knowledge, successful use of serendipitous opportunities, and even aesthetic creations (ibid.: 317).

But improvisational learning may also have unexpected harmful impacts, by drawing attention away from prior learning and related priorities, and sometimes by generating organizational conflict. Improvisation may both bear fruit and create dangers for organizations. Improvising and innovation have both a light and a dark side.

To sum up, improvisational learning underlines conditions of openness, spontaneity, variety and emergent processes that connect with Billett's and Boud's thinking. On the other hand, the outcome is broader than remaking job practices, and improvisational learning also underlines adaptation, innovative learning and organizational learning. This thinking is further elaborated in Per-Erik Ellström's theory of adaptive and developmental learning.

Ellström's theory of adaptive and developmental learning

From this perspective, employee-driven innovation is constituted by the learning and knowledge creation that take place in workplaces, and the created knowledge is seen as a learning outcome that takes place in the course of employees' everyday work (Ellström, 2010: 27).

According to this line of thought, it is important that the workplace takes the form of a seat of learning and that the learning product is not only mental changes or social processes, but also tangible results such as new ideas, new knowledge, new solutions to problems. The focus is on *process* innovations, that is, innovations that relate to production processes, working methods and work organization (ibid.: 28). In this approach, the relation between learning and innovation is quite clear: 'workplace learning is a fundamental mechanism behind practice-based innovation processes' (ibid.: 28). Learning – in terms of tangible results – produces innovation.

What, then, is the role of learning in innovation processes, according to Ellström? To answer this question, the concepts of adaptive learning and innovative learning – two complementary forms of workplace learning – are basic. Learning may be conceptualized both at a micro-level (individual learning) and at a macro-level (organizational learning). According to Ellström, 'Organizational learning can be defined as changes in organizational practices (including routines and procedures, structures, systems, technologies etc.) that are mediated through individual learning and knowledge creation' (Ellström, 2010: 37).

As regards adaptive and innovative learning, different terms are used with different and similar meanings. In the following, the terms adaptive learning, maintenance learning and reproductive learning are synonymous, and the same goes for the terms innovative learning, developmental learning and creative learning.

Botkin *et al.* (1979) distinguished between *maintenance learning* and *innovative learning*. Maintenance learning is the acquisition of fixed outlooks, methods and rules for dealing with known and recurring situations. It enhances our problem-solving ability for problems that are given. It is the type of learning designed to maintain an existing system or an established way of life. Maintenance learning tends to take for granted the values inherent in the status quo and to disregard all other values.

Ellström states that adaptive learning 'has a focus on establishing and maintaining well learned and routinized action patterns' (Ellström, 2010: 33). The focus in adaptive learning is on the subject's mastery of certain given tasks or situations – for instance, the introduction of new technology on the job – and on the refinement of task performance or, for example, of existing routines in an organization (Ellström, 2006: 44). Routinized actions are developed through formal or informal forms of learning. In this way, adaptive learning includes mastery of new tasks or situations, learning to follow certain instructions (Ellström, 2010: 33). In terms of socialization, adaptive learning may include learning in the workplace on how the individual should think and act in different situations (ibid.: 33). Ellström underlines what he calls the inbuilt duality in the concept of adaptive learning:

> In one sense, adaptive learning is about learning to handle certain tasks or to master the norms, practices and routines in an organization. In a different sense, adaptive learning is about the learning and reproduction of a prescribed order (e.g. a new policy or procedure) and, thereby, a mechanism of power and managerial control. (Ibid.: 33)

In other words, adaptive learning – with great similarity to single-loop learning – unfolds itself in a situation characterized by a high degree of specification in terms of the present situation, goals of the task and procedures to reach the goal. It is a learning process in which behaviour is cultivated:

going faster, being more precise and committing fewer faults. It is a process of routinization of action patterns. The decisive criterion for successful adaptive learning is that the task at hand can be performed rapidly and with a low percentage of error (ibid.: 33). This is one basic function of routinization, and – at first glance – this appears to be a counterpart to innovation. But the relation is not that simple.

Ellström calls attention to the issue that adaptive learning, implying routinization of action, also means that a person can master many activities without much cognitive effort, and thereby is able to reallocate attention and time from routine tasks to more creative tasks (Ellström, 2006: 46). This means that adaptive learning, even based on routinization learning, may facilitate innovative efforts: the learned routines and action patterns may work as preconditions for generation of freedom and variation of action that may support creativity and innovative learning (Ellström, 2006: 46). This is an important relation between adaptive learning and innovation.

As opposed to adaptive learning, *innovative learning* occurs in times of turbulence, change or discontinuity. It is the type of learning that can bring change, renewal, restructuring and problem reformulation.

The following are viewed as being characteristics of learning contexts that foster innovative learning: new problems, unknown situations, complex situations, turbulence, change and discontinuity (Botkin *et al.*, 1979).

Ellström states this conceptualization:

> This (developmental learning) means that there is a strong emphasis on the subjects' capacity for self-management and their preparedness to question, reflect on and, if necessary, transform established practices in the organization into new solutions or ways of working. (Ellström, 2010)

This statement indicates a close and important relationship between innovative learning and innovation processes.

It is evident that we find a high degree of correspondence between the early formulations by Botkin and Ellström's concept of developmental learning. A connecting thread throughout most of the presented theories is that the dimension of openness and closeness in the work organization is important. This observation corresponds to the empirical findings in macro and quantitative research presented earlier. Below, this question is elaborated on the micro-level, referring to the theories presented above.

Organizational conditions that facilitate innovation and learning: the role of divergent and convergent processes

Divergent and convergent processes appear to be important concepts that can bridge our conceptualization of the basic conditions and processes of innovation and learning.

Divergence–convergence – open or closed situations – are decisive factors in the context of adaptive learning, and the same goes for innovative learning. Equally, the innovative process consists of several periods of opening up and closing down, that is, of divergence and convergence (Darsø and Høyrup, 2012). Divergence is aimed at exploring, finding out, asking questions and discovering new possibilities, whereas convergence is aimed at reaching the set goal, making decisions, limiting possibilities and controlling results. Evidently, both are needed in the innovation process (Darsø and Høyrup, 2012).

Convergent processes involve a high degree of social order and ordering for reaching a prescribed and precise goal, making decisions, limiting possibilities and controlling the results of an implementation. Convergent processes, in this view, are connected to organizational routines. In other words, convergent processes imply *a reduction of complexity and variation*, a reduction of freedom to interpret the situation you are going to act in, and a reduction in the range of action possibilities. In terms of problem-solving, the problem situation is highly structured and closed in terms of specification of the situation, method for solving the problem, and outcome of the problem-solving.

In contrast, divergent processes imply *an increase in complexity and variation*, an extension of the freedom to interpret the situation you are going to act in, and an increase in the range of action possibilities. The problem situation has a low structure and is open in terms of a low degree of specification of the situation, method for solving the problem, and outcome of the situation.

From studies in innovation we know that the total innovation process consists of several periods of opening up and closing down – in other words, periods when either divergent or convergent processes are predominant. In the innovation process, the two processes may alternate, but, in broad terms, divergent processes are predominant in the first phases, including ideation and creativity, and convergent processes are predominant in later phases focusing on implementation of ideas. At first glance, it appears that the divergent processes are most important as the facilitating processes of learning and innovation.

From organizational studies and studies in workplace learning, we know that we often find divergent processes in what is called the informal organization, in informal networks, in spaces 'in between' (Solomon *et al.*, 2006), 'behind the scenes' (Brown and Duguid, 1991; Ellström, 2010; Miner *et al.*, 2001; Sundbo, 2003). These authors emphasize these kinds of social environments as very productive of innovation.

What organizational concepts can be applied to advance the conceptualization of divergent and convergent processes? Core concepts here are 'discretion', 'autonomy', 'organization logic', 'variety in work', 'open problem-solving situations' and 'chaos'.

The concept of *'organization logic'* includes logic of production and logic of development, in Ellström's terms (Ellström, 2010). 'The logic of production puts a strong emphasis on goal consensus, standardization, stability and the avoidance of uncertainty' (Ellström, 2010: 32). At the same time, this also forms the core of the convergent processes.

Conversely, 'the logic of development focuses on practice as a source of new thinking and knowledge development' (Ellström, 2010: 33). The variety/ autonomy/freedom that always exists to some degree in the performance of the work is a core element in this logic. The effort is to increase – not to reduce – variety.

Another concept within work organization theory is *variety*. The terms variety and variation may be seen as core aspects of the convergent and divergent processes. Variation concerns how the work process and tasks are conceived by the individual and how they are performed, and the individual's conception may have some degree of autonomy in relation to the formally prescribed performance and structure of the work task (Ellström, 2010: 31). On the one hand, the work process may be formally prescribed through instructions, job descriptions and management decisions. A firm (management) may strive to reduce variety through formalization by rules and instructions, or may try to increase variation. On the other hand, management decisions and guidelines are principles, general in nature, and do not turn into specifications. This observation creates an arena for interpretation by employees in terms of what to do, and how to do it, in terms of concrete actions. This paves the way for variety and some degree of autonomy in relation to explicit and formal structures of work: instructions and guidelines for work have to be perceived by different employees as the basis for performance in practice. One factor creating variety is thus constituted by different actors' different perceptions of their work situation and subsequently different performance of work processes. Problems and tasks at work may simply be interpreted differently by different players at the same time, or by the same individuals over a period of time, due to learning. This diversity of actions, deviations from 'the prescribed ways of fulfilling the job', supports the development of the divergent mode. Divergence may be produced through problem-solving that generates new solutions and actions, and when the actions of individual players in a socio-technical environment may trigger unexpected and surprising consequences. This means that, as regards innovative learning, tension, conflict and ambiguity may not be seen as threats to learning, but rather as potential triggers of developmental learning. Furthermore, research indicates that factors challenging the established system may be located outside the organization. This thinking corresponds to Chesbrough's concept of *open innovation* (Chesbrough, 2005).

Comparing adaptive and innovative learning in terms of variation, Ellström states that a basic condition for adaptive (reproductive) learning is to reduce variation (Ellström, 2010: 33). Compared with conditions for innovative learning, variation in ideas or solutions for defining and handling problems is considered to be a key factor in promoting developmental learning and practice based (employee-driven) innovations (Ellström, 2010: 36). It is the exploring of variation and diversity in thought and action that matters. In addition, a high degree of autonomy, the individual's capacity for self-management and the individual's preparedness to question and reflect on established practices, leading to new solutions or ways of working, are all important factors (Ellström, 2010: 34). However, Ellström underlines that a high degree of autonomy in performing a task is an essential but not a sufficient prerequisite for innovative learning. It is important that individuals or groups have the subjective capacity required to make use of the autonomy afforded by the job (Ellström, 2010: 36). In Billett's terms, interplay is taking place between job affordances and individual capacities. Further, risk-taking behaviour and a capacity for critical reflection, together with sufficient scope and resources for experimenting with and testing alternative ways of action in different situations, are emphasized as important conditions of innovative learning (Ellström, 2010: 34).

The concept of open and closed problem-solving situations was first elaborated by Herbst (1971). A problem-solving situation may be characterized in terms of three elements: first, point of departure for the problem-solving; second, methods or procedures applied to solve the problem; and third, the results or outcomes of the problem-solving. Each of these elements can have a high degree of pre-specification, a medium degree of specification, or a low degree of specification. This constitutes a matrix of nine prototypes of problem-solving occurrences. To make it simple, we can talk about a continuum from very closed problem-solving (point of departure is well defined, only one procedure is available to reach the goal, and you have one right outcome) to very open problem-solving (point of departure may be interpreted in many ways, a lot of routes lead to the goal, and a broad range of outcomes are all acceptable as perfect solutions). In between, you have different combinations of specifications related to the three elements.

In these terms, divergent processes may be characterized as rather open problem-solving situations, and convergent processes equally characterized as closed problem-solving situations. The concepts of adaptive and innovative learning – in the very definition of the concepts – are closely related to closed versus open problem-solving situations.

The concept of Chaos is a fourth approach to making sense of divergence versus convergence. Referring to Van Eijnatten and Putnik, chaos may be understood as 'a condition of disarray, discord, confusion, upheaval, bedlam, and utter mess arising from the complete absence of order' (Van Eijnatten and Putnik, 2004: 419). Yet, according to the chaos theory and complexity theory, chaos and order in organizations are not to be seen as opposites, but as complementary. This is underlined in the term *chaord* that indicates the simultaneous presence of both. A chaordic system is a holon with a behaviour that is both unpredictable and patterned at the same time (Van Eijnatten and Putnik, 2004). The project may thus be characterized as a chaordic system. It is remarkable that environmental dynamics such as disruption, confusion and chaos have been seen as negative factors historically, often indicating the impending demise of an organization (Goldman *et al.*, 2009: 556). When looking at the potential of the chaordic system to foster or impede different kinds of learning, it is important to note that the system may offer a high level of autonomy for individuals working within successful complex systems (Goldman *et al.*, 2009: 560). In this connection, autonomy is often stated as an important supporting factor in workplace learning (Ellström, 2010).

One finding in the above-mentioned case study by Goldman *et al.* (2009) is that the chaotic learning environment comprises important factors supporting adaptive learning. The situation ensures a great variety of diverse cases: difficult cases (problems) as well as unusual cases. Moreover, the corresponding experiences also ensure repetition of situations and problems to frame and act upon. Repetition pushes the individual to act and reflect and implies facilitation of learning including practice and *reflection*, which often leads to additional self-directed learning by, for instance, asking questions and eliciting feedback when desired. This may include self-directed activities, in terms of asking people to share knowledge, and possibly following up on self-directed activities. Intense experiences typically involve high levels of interpersonal exchange between professionals. Processes of reflection may be seen as basic to both adaptive and innovative learning, because reflection (including reflection-in-action) may correct and give stability to professional actions and, at the same time, is assumed to have a critical function – questioning the taken-for-granted and organizational routines (Eraut, 1994: 144). Further reflection may give rise to on-the-spot experiments, which is obviously not acceptable in the case of seriously injured patients in the emergency department. It may be concluded that chaos has potential for supporting adaptive as well as innovative learning situations, because of the complexity involved.

The close relation between innovative learning and the divergent mode is also supported by empirical research (Goldman *et al.*, 2009; Lam, 2005; Miner *et al.*, 2001). Ellström refers to McGrath's published empirical research indicating that 'a high degree of autonomy is likely to promote developmental (exploratory) learning' (Ellström, 2010: 36; McGrath, 2001). At the same time, we may claim that there is no one-way deterministic relationship between environment and innovation and learning processes. The sociopersonal approach – recommended by Billett – appears to be a constructive approach: individual characteristics matter in the interplay of learning and innovation in the social context. Goldman *et al.* (2009) emphasize self-direction as an important factor in learning in open environments. Elström emphasizes that 'individuals or groups must have the subjective capacity required to make use of the autonomy afforded by their jobs' (Ellström, 2010: 36). Factors such as self-confidence and occupational identity are emphasized here. Last but not least, it should be stressed that, although employees involve themselves in innovative learning and produce new experience, knowledge and solutions to problems, it is possible that these learning products are only potentially innovative movements that never get the attention or support needed to complete their journey to be coming implemented and institutionalized innovations. To fulfil this journey, the organizational culture is an important factor for paying attention to and recognizing the initiatives from employees.

In conclusion, our analysis indicates that the relation between organizational conditions for learning and learning processes is not a straightforward one. On the one hand, the concepts of convergent and divergent processes support adaptive and innovative learning, respectively. One example is that improvisational learning relies on conditions of openness, spontaneity, variety and emergent processes. On the other hand, our analytical concepts of variety, chaos and problem-solving indicate the complexity of real life organizational processes: order and chaos are not mutually exclusive; divergent processes may include elements of closed problems, and so on. Chaos may include conditions that support both adaptive and innovative processes. This complexity tends to become even more pronounced when we focus on the next basic question: how does learning produce innovation?

Production of innovation through learning

How, then, do the different forms of learning – adaptive learning, innovative learning, double-loop learning, improvisational learning, and so on – contribute to employee-driven innovation?

It appears that adaptive learning contributes to employee-driven innovation in several ways. On the individual level, as mentioned earlier, adaptive learning implies that a person develops his mastery of different activities in such a way that these activities can be executed without much cognitive effort, thereby helping the person to reallocate attention and invested time from routine tasks to more creative tasks. Further adaptive learning contributes to routinization of actions and contributes to the development of organizational routines. The same applies to single-loop learning, which is appropriate for the routine issues in the organization; it helps get the everyday job done.

This is an important fact, because routines are often prerequisites of innovation. It has been said that 'you can't improvise on nothing. You've gotta improvise on something' (Miner *et al.*, 2001: 316). One further step towards innovation is the questioning of the taken-for-granted, reflection and enquiry directed toward organizational routines. *Adaptive learning helps* to create the very preconditions of innovations, on both the individual and the organizational level.

If we are looking at the whole innovative process, Lotte Darsø– in broad terms – conceptualizes two main phases: the preject and the project. The innovation process consists of several periods of opening up and closing down, of divergence and convergence. The period of the preject is dominated by divergent processes (Darsø, 2009).

Evidently, both periods are needed in the innovation process. Creativity is dominant in the preject phase. In this phase, and in everyday working life and work practice possessing the qualities of openness/divergence, innovative learning processes predominate, contributing to innovation in different ways, as mentioned earlier. Creativity, improvisional learning and innovative learning are often found in this divergent mode. This is a common theme in published theories of learning: diversity paves the way for critical reflection, questioning the taken-for-granted, questioning the prevailing practice, and freedom for experimenting and trying out new and alternative ways of action in relation to existing problems. Experiences of new and surprising situations may also contribute to this divergent mode. Robinson and Stern describe this phenomenon as the (innovative) power of the unexpected (Robinson and Stern, 1997).

When new ideas, initiatives, actions and knowledge have been created and recognized as important for the organization, the subsequent phases of implementation and institutionalization imply that people have to adapt and learn the new practices. This is the project phase. Projects are by definition convergent. This implies adaptive learning: learning to act smoothly according to the new ideas, to act quickly and efficiently, and learning in terms of socialization. Thus, adaptive learning supports the implementation process of the renewal of the organization. These are further contributions to the ways in which adaptive learning facilitates innovation.

What, then, are the lessons learned from the different theoretical approaches elucidating the relation between learning and work? Price *et al.* state (this volume, p. 89) that there is no simple relation between innovation and work, and we can add that there is no simple relation between learning and innovation. Convergent and divergent processes in the organization support adaptive and innovative learning, and both adaptive learning and innovative learning support the production of innovation in several ways.

Conclusion

The purpose of this chapter is to unveil the relation between innovation and workplace learning in the context of the work organization. It has been demonstrated that the processes of learning and innovation are closely interwoven and constitute almost inseparable prerequisites of each other.

One important relation is that workplace learning has the potential to produce innovations of different kinds. Specifically, in terms of the remaking of work practice, process or organizational innovations are important. The concept of convergent and divergent processes emphasizes basic relations between innovation and learning – especially in terms of adaptive and innovative learning – and the organizational conditions of learning and innovation: divergent processes in the organization foster both innovative processes and innovative learning, and innovative learning has the potential to produce innovations. Yet the connection between learning of different kinds and innovation is much more complex. A prominent pattern includes the interplay of adaptive and innovative learning in relation to innovative processes. Adaptive learning produces outcomes at the individual and organizational levels that are important for innovative processes. At the individual level, adaptive learning may create a cognitive surplus for the individual that might help the individual to face innovative demands. At the organizational level, adaptive learning may produce a climate of safety and security that facilitates innovative processes. In the early phases of the complex innovative process, in the preject stage when divergent processes are dominant, innovative learning, producing innovation, is fostered. During the project stage, dominated by convergent processes, when creative and useful ideas have to be taken up and implemented by a broader range of employees, adaptive learning may support this process. In this way, adaptive and innovative learning – and related learning processes such as improvisation – are basic mechanisms that facilitate innovation in different ways and in different phases of the innovation process. These interplays of processes are important learning mechanisms behind employee-driven innovation.

References

Alter, N. (Ed.) (2000). *L'innovation ordinaire*. Paris: Presses Universitaires de France.
Argyris, C. and Schön, D. (1978). *Organisational learning: A theory of action perspective*. Reading, Massachusetts: Addison Wesley.
Billett, S., Smith, R. and Barker, M. (2005). 'Understanding work, learning and the remaking of cultural practices'. *Studies in Continuing Education*, 27(3), 219–237.
Botkin, J., Elmandjra, M. and Malitza, M. (1979). *No Limits to Learning*. Oxford: Pergamon.
Brown, J.S. and Duguid, P. (1991). 'Organizational learning and communities-of-practice: Toward a unified view of working, learning and innovation'. *Organization Science*, February, 2(1), 40–57.
CERI/STI (2010). *Innovative workplaces learning organizations and innovation*. OECD.
Chesbrough, H.W. (2005). *Open Innovation. The New Imperative for creating and Profiting from Technology*. Boston, Massachusetts: Harvard Business School Press.
Darsø, L. (2009). *Innovation in the Making*. Copenhagen: Samfundslitteratur.
Darsø, L. and Høyrup, S. (2012). 'Developing a framework for innovation and learning in the workplace'. In Melkas, H. and Harmaakorpi, V. (Eds) *Practise-based*

Innovation: Insights, Applications and Policy Implications. Berlin, Heidelberg: Springer-Verlag.
Drucker, P.F. (1987). 'Social Innovation – Managements New Dimension'. *Long Range Planning*, 20(6), 29–34.
Ellström, P.-E.(2006). 'The meaning and role of reflection in informal learning at work'. In Boud, D., Cressey, P. and Docherty, P. *Productive Reflection at work*. London and New York: Routledge.
Ellström, P.-E. (2010). 'Practice-based innovation: a learning perspective'. *Journal of Workplace Learning*, 22(1/2), 27–40.
Eraut, M. (1994). *Developing Professional Knowledge and Competence*. New York: Routledge Falmer.
Evans, K., Hodkinson, P., Rainbird, H. and Unwin, L. (2006). *Improving Workplace Learning*. London, New York: Routledge.
Fenwick, T.J. (2003). 'Innovation: Examining workplace learning in new enterprises'. *Journal of Workplace Learning*, 15(3), 123–132.
Goldman, Plack et al. (2009). 'Learning in a Chaotic Environment'. *Journal of Workplace Learning*, 21(7), 555–574.
Hargadon, A. (2003). *How Breakthroughs Happen. The Surprising Truth How Companies Innovate*. Boston, MA: Harvard Business School Press.
Herbst, P.G. (1971). Struktureringavkunnskapogutformningavutdanningsorgani sasjoner' (Organization of knowledge and design of educational institutions). Nordisk Forum. *Tidsskrift for universitets- & forskningspolitik*. 6, 171–188. Copenhagen: Munksgaard.
Høyrup, S. (2010). 'Employee-driven innovation and workplace learning: basic concepts, approaches and themes'. *Transfer*, Summer, 16(2), 143–154.
Jensen, M., Johnson, B., Lorenz, E. and Lundvall, B.-A. (2007). 'Forms of knowledge and modes of innovation'. *Research Policy*, 36(5), 680–693.
Kesting, P. and Ulhøi, J.P. (2008). *Employee Driven Innovation: The Discovery of the Hidden Treasure*. Draft, August.
Kesting, P. and Ulhøi, J.P. (2010). 'Employee-Driven Innovation: extending the license to foster innovation'. *Management Decision*, 48(1), 65–84.
Lam, A. (2005). 'Organizational innovation'. In Fagerberg, J., Mowery, D.C. and Nelson, R.R. (Eds.) *The Oxford Handbook of Innovation*, Oxford: Oxford University Press, 115–147.
LO (The Danish Confederation of Trade Unions) (September 2008) *Employee-Driven Innovation: Improving Economic Performances and Job Satisfaction*. Provisional printed version, Copenhagen.
Lorenz, E. and Valeyre, A. (2005). 'Organizational Innovation, HRM and Labour Market Structure: A comparison of the EU15', *Journal of Industrial Relations*, 47, 424–442.
McGrath, R.G. (2001). 'Exploring learning, innovative capacity, and managerial oversight', *Academy of Management Journal*, 44(1), 118–131.
Miner, A.S, Bassoff, P. and Moorman, C. (2001). 'Organizational Improvisation and Learning: A Field study'. *Administrative Science Quarterly*, 46(2), 304–337.
Robinson, A.G. and Stern, S. (1997). *Corporate Creativity. How Innovation and Improvement Actually Happen*. San Francisco: Berrett-Koehler Publishers, Inc.
Schatzki, T.R. (2002). *The Site of the Social: A Philosophical Account of the Constitution of Social Life and Change*. Pennsylvania State University Press.
Senge, P.M. (1990). *The Fifth Discipline: The Art and Practice of the Learning Organization*. London:Random House.

Schumpeter, J.A. (1934). *The Theory of Economic Development*. Cambridge, MA: Harvard University Press.
Solomon, N.V., Boud, D.J. and Rooney, D.L. (2006).'The in-between: exposing everyday learning at work'. *International Journal of Lifelong Learning*, 25(1), 3–15.
Sundbo, J. (2003). 'Innovation and strategic reflexivity: an evolutionary approach applied to services'. In Shavinina, L.V. (Ed.) *The International Handbook on Innovation* 87–114. Boston: Elsevier Science Ltd.
Teglborg, A.-C. (2010a). 'Modes of approach to employee-driven innovation in France: an empirical study'. *Transfer*, Summer, 16(2), 211–226.
Teglborg, A.-C. (2010b). 'Les dispositifs d'innovation participatives. Une reconception reflexive á l'usage'. *Ecole doctorale Paris I Sorbonne*. IAE de Paris – HEC Paris, Paris. (PhD thesis.)
Van Eijnatten, F. and Putnik, G.D. (2004). 'Chaos, complexity, learning and the learning organization. Towards a chaordic enterprise'. *The Learning Organization*, 11(6), 418–429.

2
Employee-Driven Innovation: Operating in a Chiaroscuro

Ann-Charlotte Teglborg, Renaud Redien-Collot, Maria Bonnafous-Boucher and Céline Viala

As of 2010, six of the ten most innovative companies in the world were American, two were Japanese and one was Korean. None were European. Three years before, in 2007, the situation in the Old World had been hardly any brighter; according to the European Innovation Scoreboard, the innovation index stood at 0.73 in Sweden, 0.59 in Germany and 0.47 in France. In the same year, the World Intellectual Property Organization listed the number of patents lodged by country. Sweden's 1,287, Germany's 17,739 and France's 12,112 were dwarfed by the USA's 157,283 and Japan's 164,954. It seems undeniable that many Western countries are still a long way behind and that closing the innovation gap remains a kind of quest for the Holy Grail in terms of competitiveness and employment. However, on closer inspection, the keys to Paradise are more like a safety device both for national economies and for the men and women who implement innovations on a day-to-day level.

In terms of encouraging innovation, firms are paying ever greater attention to locating and exploiting its many and varied potential sources. Alongside traditional approaches such as Technology-Driven Innovation and User-Driven Innovation, new perspectives are being explored, with Open Innovation, Design-Driven Innovation and Employee-Driven Innovation proving to be particularly popular. Indeed, it is obvious that all drivers (sources) of innovation must be used, and employee-driven innovation is a very important and effective resource for innovation (Høyrup, 2010). As is clearly demonstrated by a close reading of the industrial history of the nineteenth century – studded as it is with examples of often decisive innovations developed by workers – the informal participation of employees in the process of innovation is not a new phenomenon (Alter, 2000). In the current socio-economic context, informal employee participation often tends to be transformed into open and intentional participation in the firm's drive towards innovation. Moreover, employee-driven innovation is defined by LO as the active, systematic contribution of employees to innovation (LO, 2007).

This notion of employee participation contributes to the formation of an ideology which, on occasion, casts employee-driven innovation (EDI) as a miraculous panacea. Indeed, the task of creating a consensus about the usefulness of EDI presents little difficulty in a context in which all the stakeholders – employees, the firm itself, the shareholders and the socio-economic territory – appear to benefit from it.

The objective of this chapter is to reappraise the irenic image of employee-driven innovation and reveal the chiaroscuro characterizing management under pressure, the veritable driving force of the organization of labour in the twenty-first century. Without calling into question the urgency with which Europe needs to acquire the means of rivalling the growth of other regions in the world, and without disputing the notion that innovation stimulates competitiveness, we point out that the paradoxes inherent in innovation and the task of managing it continue to generate contrasting outcomes. As Vaughn (1999) suggests, an emphasis on competition interferes with the delicate balance of social relations providing the source of innovation (Latour, 1991). A focus on competition leads to innovation processes being formalized in a more radical manner (Letiche, 2009, Vaughn, 1999), thus restricting opportunities for dialogue with oneself and others. Therein lies one of the paradoxes of EDI; designed, amongst other things, to encourage dialogue about potential innovations, joint projects and practices, when EDI is overly formalized it often succeeds in doing the opposite. Nevertheless, formalization is necessary; in its absence, the implementation of ideas is either ephemeral or ineffectual (Letiche, 2009; Serres and Latour, 1995), obliging actors to start from scratch every time they do something new. With regard to EDI's contrasting outcomes and to the tensions of which it is a driving force, how can the organization and management ensure a balance between the imperatives of formalization and the effective interactions of network actors? More generally, are actors aware of the chiaroscuro projected by EDI onto innovation management, and what can they learn from it?

For firms, EDI outcomes are persuasive. It is perhaps in this area that we should attempt to discern how, over time, linking innovation management to outcomes has encouraged both positive dynamics, generating an increasing number of fruitful employee initiatives, and repressive dynamics, characterized by inhibition and a lack of motivation.

In the first part of this article, we outline our methodological approaches, before moving on to present two extreme cases selected from a sample of 20 studied in partnership with the Innovacteurs Association between 2007 and 2009 (Teglborg, 2010). These two cases highlight the way in which firms ensure that EDI management and the outcomes generated by it are closely linked. In the second part of the article, we will see how this encourages the subjectivation of actors at both individual and inter-individual levels by boosting the entrepreneurial skills of employees and teams at the collective level and fostering a degree of complicity between actors in terms of the development of an original approach to innovation. In both cases, the EDI approaches foster competitive advantages and develop a learning environment for employees. In the third part of the article, we will see how EDI can generate organizational and managerial pressure and, in the medium term, can cause a certain number of employees to lose interest. In the fourth part of the article, we will discuss how firms involved in EDI sometimes deal with contrasting outcomes and attempt to establish a balance.

An empirical analysis grounded on two French industrial monographs

For firms, the outcomes of EDI are persuasive. Nevertheless, initially, referring to two case studies, we would like to point out that such outcomes are accompanied by a certain degree of strategic and managerial formalization which casts a shadow over the intention to unleash the creative potential of employees. Vaughn (1999) underlines that, in all processes of organizational innovation, there is a time for the informal development of potential via chance meetings and associations, and a time for imposing a greater degree of formalization, which is both necessary and increasingly constrictive. Similarly, we will see from an analysis of the two case studies that, by involving employees in the fact that the outcomes obtained correspond to a certain type of managerial approach, the firm's middle management seeks to legitimize an ever-increasing degree of formalization in EDI processes.

Methodology

The approach used, which is informed by a desire to explore certain specific phenomena and suggest new perspectives rather than to test theoretical hypotheses (Eisenhardt and Graebner, 2007), is derived from a holistic multi-case study (Yin, 1989) of 20 firms conducted in partnership with the Innovacteurs Association and analysed in a doctoral thesis (Teglborg, 2010). The 20 individual cases were selected because they each emphasize a different facet of the EDI phenomenon and also because they represent good practices. The primary sample of 20 case studies is, in effect, heterogeneous to the extent that it includes firms of various sizes, with different organizational approaches, active in different economic sectors. The research design thus encourages the type of theoretical replication defined by Yin (1989).

The documentation provided by each company describing its individual approach to EDI, often in great detail, was subjected to analysis. The EDI manager or managing director was then interviewed, after which a group interview was conducted with a sample of eight to twelve people involved in the EDI apparatus. Finally, four individual interviews with employee–innovators were carried out (with the exception of a single case study).

In this chapter, we have chosen to focus on two case studies selected from the primary sample of 20. The two cases can be considered *extreme* in that they are particularly revealing of different facets of the phenomenon under study (Yin, 1989). According to Siggelkow (2007), cases can be used as a source of either *inspiration* or *illustration*. For Siggelkow, 'In fact, it is often desirable to choose a particular organization precisely because it is very special in the sense of allowing one to gain certain insights that other organizations would not be able to provide' (Siggelkow, 2007). As well as providing inspiration, the two cases have an illustrative potential and offer the possibility of fleshing out the conceptual constructs outlined in this chapter.

The two companies analysed have both been involved in employee-driven innovation for over ten years. At Solvay and Favi, employee-driven innovation is based, on the one hand, on a pronounced orientation toward the market – or, in other words, on a focus on clients – and, on the other hand, on clearly framed strategic choices shared with the employees. The choice of two corporate profiles, the first a group, the second an SME, is deliberate; when it is orientated toward organizational entrepreneurship, employee-driven innovation can be adapted to a wide range of structures and sectors.

Two entrepreneurial EDI innovations

The Solvay-Tavaux case: a series of decisive, polymorphous innovations over a ten-year period

Located in the Jura region of eastern France, Solvay's Tavaux factory specializes in the production of chemical products and plastics for industrial companies. The *Innovation*[1] approach was implemented at the factory, which employs 1,500 people, in 1999. Solvay-Tavaux is regarded as one of the most innovative sites in the Solvay group, with a participation rate of 55.4 per cent of employees in 2008. The firm's objective is to persuade all its personnel to be innovative. Both incremental and radical innovation projects are encouraged. Thus, projects aimed at improving safety, processes and customer service, and ideas for finding new markets for the chemical products and plastics produced and sold by the factory to industrialists are suggested and supported, thereby enabling the factory to increase its volume sales.

Finding what you're not looking for: serendipity at the service of radical innovation

In EDI processes, employees often end up finding what they were not originally looking for. The term 'serendipity' is applied to unexpected discoveries made during the course of research projects initially focusing on different objectives. Amongst the radical ideas that have emerged in such processes, the use of the chemical component 143a in freezer coolants is particularly interesting. Chemical component 143a is a sub-product of the PVDF monomer, a specific kind of monomer renowned for its technical characteristics. The component was initially treated as a waste product. Its characteristics as a coolant were known to people operating on the ground; an installer suggested that 143a could be reused in mixtures for coolants and replace components which damaged the ozone layer. After conclusive tests carried out at the Tavaux Centre for Applied Studies and Research, the product was put on the market in coolant mixtures, thus creating a new market. The market took off and a chemical reactor dedicated to producing the high added value product, not long since treated as mere waste, was built at the Tavaux site. Meanwhile, an employee at the company came up with the original idea of developing PVC tiles and selling them to a manufacturer of traditional tiles. The new PVC products enabled Solvay to develop a new market for its products.

Managing employee-driven innovation

The employee-driven innovation dynamic encourages all employees to suggest and implement new ideas with a view to ensuring that the firm's strategic ambitions bear fruit.

Challenges are regularly organized with a view to eliciting ideas about specific subjects. For example, a challenge on the topic of water is held in partnership with the Besançon University Institute of Technology. Another challenge has been developed to encourage all employees to contribute to the manufacture of a new finished product, while a third challenge focuses on energy savings in a factory which consumes as much electricity as the city of Lyon. The best ideas are selected and developed: in 2008, out of 2,185 ideas suggested, 633 were implemented.

In addition to spontaneously suggested ideas and the organization of challenges, brainstorming sessions on key subjects are held involving various members of staff and management, including line operators, experts, heads of department and engineers. One of the departments working on fluorides holds brainstorming sessions on re-engineering part of the fluoride process. Such sessions have made it possible to reappraise the chemical plant entirely and save 4 million euros by suggesting ideas for reducing overheads and transforming what was formerly thought of as a waste product into a product that can be sold to Solvay's clients.

And this is not all that happens at the Tavaux site. An efficient system for collecting, processing and developing ideas has been developed: the IT platform, *Innoplace*.

Decisive results in terms of innovation, organizational entrepreneurship and entrepreneurial orientation.

The approach generates significant profits, ranging from 2,200 to 4,800 euros per year per employee. In 2008, 2,185 ideas were suggested by the

1,500 employees. Of those ideas, 633 were implemented, accounting for a total of 400,689 euros of profits for Solvay. In addition to radical innovations creating new markets, a multitude of incremental innovations also serve to reduce costs.

As with many other firms involved in the field, Solvay-Tavaux's approach to communication about EDI changed over the years. At the outset, the company focused on the results of the previous year, with the aim of fostering an interest amongst its employees in upcoming EDI projects in the following year. Gradually, within the framework of Innovacteur challenges, Solvay-Tavaux highlighted and rewarded a number of good practices. In the late 1990s, the firm started to focus on tools (Innoplace) and formal processes supporting EDI. These processes were also designed to impress shareholders, who could not help but appreciate the rigour with which they had been implemented. But they also targeted employees, with a view to convincing them that it was necessary to formalize EDI processes in order to more effectively integrate their initiatives into the firm's strategy. According to the firm's middle management, this alignment of employee practices, innovation processes and the diffusion of innovations was clearly the way to go.

The case of Favi: from long-term employee-driven innovation to an embryonic version of organizational entrepreneurship

Favi is an SME specializing in the injection of cuprous alloys. The firm's reputation was originally built on the production of water counters. It gradually diversified thanks to an innovation approach encouraging the participation of all of its employees. The firm's profits are twice the industry average. Over the last 15 years, 15–20 per cent of the firm's turnover has been reinvested in R&D, a figure that compares favourably with the industry average of 2–3 per cent. Favi has thus been able to produce a slew of technological innovations, including injected brass siphons, gearbox forks and high conductivity motors.

An eloquent example: the invention of the high conductibility rotor

The example of the high conductibility rotor is eloquent. In 1997, D., a salesman at Favi, having obtained an order for water counters from a major European industrial company, talked freely with his client about the latter's projects. He understood that the industrialist was looking for a way of improving conductivity to reduce the energy consumption of electric rotor engines. When he got back to France, D. went to the factory to organize the delivery of the order and noticed a vat full of copper in fusion. Perhaps copper could be used to do something for the client, he thought. But one of the factory's technicians told him that his idea was not realistic as it would require heating copper to 1,000 degrees Centigrade, a feat that had never been achieved before. A year later, after having had his idea rejected by every department in the company, D. was able to convince the managing director that an attempt should be made to implement his idea. After ten years of controversy, the copper electric rotor engine was produced and commercialized. Workers most specialized in the fusion of copper worked with a research team from an engineering school. Without a mysterious alliance between the tacit expertise of the SME's employees, who were used to working with copper, and the scientific knowledge of the engineers, the achievement of heating copper to 1,000 degrees Centigrade would have been unthinkable. The conductivity of the new rotor is superior to previous products and provides an energy saving of 3 per cent. With its many potential uses, the rotor is a vital product for the company, which is faced with stiff competition from Asian firms in the water counter market and a significant decline in demand for gear box forks.[2]

The rotor, used in high yield engines, pumps, compressors, traction motors and electrical power steering systems, generates substantial profits for Favi.

The Favi management system: a combination of continuous, hybrid and participatory innovations

Favi's profitability is based on the continuous management of employee-driven innovation. In 1997, workers, mini-factory leaders, sales people and directors got together to work on a project entitled 'How can we provide a framework for living for 100 families in Flavicourt in 2027?' They kick-started a process designed to find a new product that could one day replace those that had made Favi so successful. 'To last' is, in effect, Favi's *raison d'être*. This approach was suggested to the Managing Director of the Picardy-based SME by the oak tree which stands outside the factory lobby. Favi respects its clients and attempts 'to understand their needs in the context of a deep respect for the land of our ancestors'. Favi encourages sustainable development and seeks a 'balance between man, nature and economics'. Money is seen as 'oxygen for the company'.[3] Thanks to this prospective approach, Favi encourages entrepreneurial values and attitudes designed to prompt sustainable growth. Echoing this attitude, employees are encouraged to develop a commitment to innovation beyond their normal duties.

Major transversal projects encourage a convergence between innovative behaviours and entrepreneurial attitudes. Favi also stays in close contact with its clients in terms of the design of new pieces or products involving the R&D department, sales agents and operators. These last are closely involved in monitoring the firm's activities; over the course of the last ten years, 60 of them have visited factories in Poland, the USA and Japan.

The mini-factory: organizing work processes with a view to encouraging autonomy

The Favi system is based on a series of mini-factory units, each one of which is dedicated to a particular client. Each mini-factory has a leader responsible for a group of between 20–30 operators. Under the authority of their leader, mini-factories define their own objectives and organize everything necessary to meet the needs of the clients for whom they are individually responsible. Work on improving existing products is carried out in close collaboration with the client. Each mini-factory has a 'sponsor' who functions as a sales agent on its behalf. The sales agent also deals with suppliers and works closely with the mini-factory operators with a view to improving existing products and suggesting ideas on how to improve processes. This autonomy enables employees to be more proactive, since their suggestions are subject to only a few stages of validation: they can therefore be confident that the changes they suggest will be implemented.

Solvay and Favi: EDI's positive dynamic

Thanks to the Solvay and Favi cases, we can see how EDI motivates staff, making it possible to attract and retain talented employees and to manage change and complexity. At the level of the firm, it enables the people to seize opportunities, to generate innovations and develop business. It encourages flexibility, productivity and growth. Developed at Solvay, the chemical component 143a, which was a waste product, is now used for PVC tiles. EDI has led to technological innovations, which give Favi a competitive advantage (Favi). The increased conductivity of the new rotor made possible a 3 per cent energy saving (Favi). EDI also contributes to economic development and helps improve the social climate by reducing stress levels amongst employees. The values it generates (autonomy, added responsibility and communication) enable companies to reduce frustrations deriving from a lack of control or interest on the part of employees concerning their functions, as well as from factors such as poor communication. In terms of shareholders, EDI boosts profitability by increasing the capacity to constantly improve processes, products and services, to seize opportunities, to innovate, and to mobilize staff. At Solvay, where profits range from 2,200 euros to 4,800 euros per employee, 4 million euros were generated by suggesting ideas for reducing overheads.

Expected benefits: competitive advantages and a learning environment for employees

More generally, an analysis of the Solvay-Tavaux and Favi cases shows that employee-driven innovation, especially when oriented towards organizational entrepreneurship, reveals the way in which actors undergo original forms of subjectivation. Processes of subjectivation are manifested in two ways. At the collective level, EDI encourages actors to work together to develop original approaches to innovation, or, in other words, to use collective hybrid approaches, rather than applying radically new perspectives. Moreover, at the individual and inter-individual levels, EDI helps to develop the entrepreneurial skills of both teams and individuals. This is demonstrated by a certain number of entrepreneurial initiatives. These two original forms of subjectivation, which encourage actor network links and innovation, underpin two forms of learning.

Employee-driven innovation encourages employees to explore an alternative form of innovation, known as *hybrid innovation*, which encompasses both radical and incremental breakthroughs. This approach itself encourages actors to work together and to form social bonds. Radical innovation 'profoundly modifies habitual approaches to the development and cost of products. It generally implies a recourse to new skills, especially when it calls upon a generic and previously unused form of technology' (Broustail and Fréry, 1993). Radical innovations create new markets and encourage new activities within the firm (internal corporate venturing), as witnessed by examples of innovation such as the high conductibility electric rotor engine developed by Favi and the use of component 143a in coolants at Solvay-Tavaux. As well as radical innovations, EDI encourages, in both of these cases, the emergence of a multitude of incremental innovations which 'lead to a gradual improvement in standards, both in terms of products and costs, which does not require new forms of expertise' (Broustail and Fréry, 1993). Workers at Favi regularly suggest ways in which the production process can be simplified, while employees at Solvay have helped develop innovations such as the sealant collar with pinhole derivation which makes it possible to keep the plant functioning during maintenance work. Similarly, brainstorming sessions served to partially reappraise operational approaches at the Tavaux chemical plant, thus saving 4 million euros. Innovations suggested by employees whose primary role does not involve innovation have enabled Favi and Solvay to boost their competitive advantage, either by reducing costs or focusing on differentiation by developing Schumpeterian innovation. This hybrid approach to innovation integrates actors into a process of incremental innovations, thereby establishing a sustained dialogue, a high degree of complicity and, in the end, a certain shared sense of pride.

Moreover, the innovations developed within the framework of the EDI approaches applied at Favi and Solvay demonstrate that those approaches have encouraged a form of Entrepreneurial Orientation (EO) amongst employees (see table 2.1). EDI encourages employees to take responsibility and develop proactive attitudes, thus contributing to their sense of accomplishment and professional fulfilment, while at the same time developing their skills and improving their employability. For example, Favi mini-factory units give autonomy to employees who become more proactive and organize the units as if they were their own businesses. Similarly, the Favi challenge providing a framework for living for 100 families in Flavicourt in 2027 has made it possible to better articulate individual and entrepreneurial projects with the firm's overall objectives. In this sense, those approaches tend towards organizational entrepreneurship. As is demonstrated in the table below,

Table 2.1 Favi and Solvay-Tavaux satisfy the three variables of entrepreneurial orientation

	Favi	Solvay-Tavaux
Product-market innovation	• Radical innovation: the rotor • Incremental innovations: cost reductions	• Radical innovation: 143a • Incremental innovations encouraging cost reductions
Proactive decision-making	• Management of mini-factories • Innovation based on client needs	Management of ideas and innovation facilitated by Innoplace
Risk-taking	Strong strategic mobilization in favour of innovation	Capacity to launch continuous waves of strategic renewal

organizational entrepreneurship is manifested in the three variables highlighted by Basso Fayolle, Bouchard (2009).

There is an advantage to developing employee-driven innovation by intensifying entrepreneurial orientation: by regularly measuring the benefits of the two kinds of innovation – incremental and radical – the organization as a whole succeeds in establishing continuity in the way in which the two different types of innovation are respectively implemented. Directors, middle management and employees all focus equally on incremental and radical innovations and risks are taken by simplifying validation and decision-making processes in order to ensure that employees continue to innovate.

In terms of employees, it creates a healthy working environment which encourages employee learning. In effect, through the complicity encouraged by the gradual emergence of hybrid innovations and by the entrepreneurial approach, EDI engenders two forms of learning: reproductive and creative (Ellström, 2006). The notion of adaptive or reproductive learning has a focus on a subject's adjustment to and mastery of certain specific tasks or situations. This is in contrast with developmental learning, where the focus is on transforming rather than reproducing a prevailing situation. This means there is emphasis on exploring and questioning existing conditions, solving ambiguous problems, and developing new solutions.

Employee-driven innovation at the heart of four major tensions

First, while providing a vehicle for multiple benefits, employee-driven innovation has, over the years, become associated with an increasingly formalized organizational and managerial apparatus covering middle management. While formalization is necessary, its adverse effects should still nevertheless be examined. Four major tensions can be observed. The first tension manifests itself at the organizational level, confronting actors with the contrast between the autonomy provided by EDI and the mainstream advocates of the firm's traditional approaches. Although inevitable, this tension can be assuaged by the introduction of a process of *intéressement* as suggested by Actor Network Theory. Second, employee-driven innovation Runs the risk of being undermined by the way in which it is managed. This tension is endogenous; formalizing EDI management engenders restrictions, and, therefore, inhibitions. Third, from a strategic point of view, there may be a lack of articulation between the various sources of innovation. Lastly, at a more personal level, disappointment and potential loss of interest on the part of employee–innovators constitutes a fourth potential problem. In effect, without being able to establish a systematic correlation, it is possible to identify signs of disappointment and demotivation resulting from one of the four tensions listed above.

Employee-driven innovation faced with organizational norms

The innovators who developed the electric rotor engine were able to implement a developmental creative learning process. Indeed, the electric rotor engine innovation required an emphasis on exploring and questioning existing conditions, solving ambiguous problems, and developing new solutions (Ellström, 2006). This learning process led the innovators to question existing norms and organizational knowledge, notably the assumption that it was impossible to heat copper to a temperature of 1,000 degrees Centigrade.

Since it calls into question the norms governing the diffusion of information, production processes, cooperation and decision-making, innovation often creates friction with systems of organizational control (Alter, 2000). In this regard, the deployment of employee-driven innovation is aiming to limit friction between innovation and organizational control.

But, in certain cases, the practice of EDI does not stop initiators of ideas from coming up against the wall of organizational norms. When the deployment of employee-driven innovation enables employees to suggest radical innovations, tensions between the employee–entrepreneur and organizational norms are very strong. The project initiator is obliged to implement a socio-discursive tactic similar to the one developed by the inventor of the electric rotor engine at Favi. Only his ability to persuade Favi's technical managers and Managing Director that it was possible to heat copper to 1,000°C and then to encourage the development of a number of alliances – with the engineering school, specialist workers and technical managers, among others – can explain the successful passage from invention to innovation which gave rise to the electric rotor engine.

The Favi case study shows that, in order to implement this radical innovation, the employee had to deploy a socio-discursive approach which, while ensuring that alliances were maintained and strengthened, made it possible to define the stages at which the members of the managerial hierarchy and the guarantors of the company's productive norms could be persuaded of the idea's legitimacy and encouraged to support it. But, what of those who are unable to elaborate persuasive approaches of this kind?

Without particular attention, radical innovations suggested by employees run a significant risk of being undermined by organizational norms. In order to be successful, some innovative projects are obliged to deploy an *intéressement* process of the kind described in Actor Network Theory. The *intéressement* process consists of gaining the support of numerous allies, thereby rendering the proposition contained in the management apparatus indispensable (Akrich *et al.*, 1988). It combines acts of seduction with acts of persuasion and even the creation of new constraints. With such a process of *intéressement*, most innovations which are accepted in EDI approaches improve what already exists without calling it into question and are mainly focused on adaptive learning (Ellström, 2006).

Employee-driven innovation is thus as fragile as it is promising. This is doubtless what encourages French companies to attempt to *manage* EDI in order, not without ambivalence, to reduce its random character.

Nevertheless, such attempts can sometimes produce negative results.

Employee-driven innovation undermined by the way in which it is managed

Three motives for designing an employee-driven innovation apparatus

Attempts to manage employee-driven innovation are characterized by the elaboration of EDI apparatuses with three levels of justification (Teglborg, 2010). First, in the development of an employee-driven innovation apparatus, the management clearly encourages all the firm's employees to participate in the adventure of innovation. The objective here is to do away with certain inhibitions on the part of employees used to seeing their initiatives curtailed by organizational norms. This kind of managerial support is one of the conditions favouring practice-based innovation (Ellström, 2006). For Ellström (2006), innovation processes are conceptualized as learning processes. Furthermore, one of the functions of the employee-driven innovation management apparatus is to ensure that employees share any interesting ideas they develop. Finally, when apparatuses are based on 'technical substrates', a 'management philosophy' and a 'simplified representation of the organization' (Hatchuel and Weil, 1992), they imply a desire to make employee-driven innovation a permanent feature of the firm. Nevertheless, the fact that all the firms within the sample (Teglborg, 2010) wanted to manage employee-driven innovation implies a certain degree of risk. In effect, EDI apparatuses are underpinned by the diffusion model of innovation (Teglborg, 2010), a model which fails to take into account the complex trajectory of new ideas. Moreover, the employee-driven innovation apparatuses implemented do not always prove to be faithful servants (Moisdon, 2003; Teglborg, 2010), and approaches based on calculated objectives sometimes turn out to be a mirage (Supiot, 2010; Teglborg, 2010).

The diffusion model as a postulate of employee-driven innovation apparatuses
Attempts to manage employee-driven innovation by implementing specific apparatuses are informed by the myth of the separation between the social and the technical (Callon, 1994; Teglborg, 2010). This myth distinguishes the design phase from the diffusion phase and the content of the innovation from its context. In this perspective, innovation is considered as a *linear series* of phases relatively independent of one another (Callon, 1994; Nieuwenhuis, 2002). The original idea gives rise to plans, layouts and prototypes confined to the technical field. Invention only breaks free of this sphere and enters the social field when all the tests are positive. Invention becomes innovation when users begin to understand it, or, in other words, when the new product meets the demand for which it was designed. The innovation then has to be diffused and its market expanded (Akrick *et al.*, 1988). According to the *diffusion model*, products, considered according to their own properties, conquer their markets purely by means of demonstration (Callon, 1994). Thus, in the diffusion model, innovation is presented as a phase in which technical problems are solved followed by an introduction to the market *where the innovation spreads by a process of contagion thanks to its intrinsic qualities* (Callon, 1994).

The real story of innovations is, in effect, entirely different, following a serpentine trajectory over the course of which technical content and the social environment evolve conjointly (Callon, 1994). Since each element is dependent on all the others, innovation continually takes on different configurations. Successful innovation changes continually by gradually integrating the expectations and interests of different actors, as in the case of the electric rotor engine at Favi, in which potential clients and qualified workers held a dialogue with researchers who integrated the knowledge and expectations of the firm's various internal and external partners. Thus, a successful innovation integrates into its very design the social environment necessary for it to function, or, in other words, the expectations and interests of the actors concerned by the innovation, be they internal or external to the firm. This conception developed by ANT is close to Nieuwenhuis's (2002) interactive model of innovative processes.

Indeed, employee-driven innovation managers have emphasized the importance of the firm's decision-makers and experts providing employee-entrepreneurs with aid and advice, but have failed to systematically integrate the expectations of future users of the innovation, be those users internal (other employees, management) or external (clients, suppliers, institutions). There are, nevertheless, a number of exceptions. At Solvay in particular, four years ago, links began to be established between employee-driven innovation and open innovation. However, the process was a cautious one, due to the fact that organizers of the firm's employee-driven innovation apparatus became aware of the highly serpentine character of innovation, or, in other words, realized that its technical contents and social environment evolve together (Callon, 1994). However, the advocates of employee-driven innovation have not yet recognized that it must accommodate compromises with regard to the interests of various actors. Tacitly, in certain departments, the *interest model* holds sway. According to this model, success is based on the capacity to make numerous alliances. In order to do so, actors integrate the demands, expectations and observations of the parties involved in the apparatus. Nevertheless, this approach is not common, and, in most cases, employee-driven innovation projects are handled by a restricted number of decision-makers and experts who do not take into account the expectations and interests of future users of the innovation.

Employee-driven innovation apparatuses: servants who are frequently unfaithful

In the hope of managing employee-driven innovation, organizers implement management apparatuses which place obstacles in the way of sharing ideas. The use of the Ideas Box at Solvay-Tavaux, while useful in terms of recording suggestions, nevertheless encourages a type of codified exchange and limits the effective interaction and co-construction of such suggestions. Perhaps this lack of effective interaction is compensated for by the type of brainstorming sessions introduced at Tavaux a few years ago, in which ideas were thoroughly sounded out by all those present. Conversely, the degree of formalization of challenges and selection processes organized with a view to implementing ideas can hinder interaction to a large extent. Ideas retained in challenges may be subject to overly precise criteria (at that stage in their development) or they may be too closely associated with organizational strategy to retain their open character and ensure that the employees responsible for them maintain the degree of motivation required to see them through. Organizations do not ask enough questions about the fact that employee-driven innovation encourages both innovative behaviours and a high degree of conformity.

The organizers of EDI challenges ask few questions about the codes governing the exploratory phases of new ideas. Not many questions were asked in either case about how to encourage workplace learning, even though it is indispensable to EDI (Ellström, 2006). Lastly, when ideas are undergoing a process of implementation, the role of experts should be fully understood: they can induce employees to submit to their authority, thus encouraging them to seek *a posteriori* approbation from a single type of actor, or, in other words, to participate in an extremely monological process of innovation.

Designed to serve the authorities who have implemented them, the apparatuses function as unpredictable servants, often all the more dangerous in that they are partially invisible (Berry, 1983). Observers can get the feeling that they are faced with a quartet of tools which reveal themselves during use to be, at best, provisional crutches which rapidly become obsolete (Boussard and Maugeri, 2003) and, at worst, supposed aids which, in reality, lead an almost clandestine existence (Moisdon, 1997); this can sometimes be counterproductive in that they cause adverse effects (Moisdon, 2003).

In 2007, a large French company implemented an employee-driven innovation system with a view to dynamizing and transforming the enterprise. The designers of the approach did not want to give money to the firm's employees. Consequently, a points system was introduced with a view to rewarding innovative ideas with gifts. When an idea was accepted by one of the firm's experts, the employee had to draw up a business plan following a five-stage procedure. At each stage, the author of the plan received extra points; for ideas put into practice, the employee received points worth up to 50 euros. However, the approach had a number of drawbacks. First, no distinction was made between simple and complex ideas. Furthermore, undue emphasis was placed on the conceptual development of ideas to the detriment of their actual implementation. The same rewards were given for ideas that had been accepted but not implemented as for ideas which were implemented. In fact, results indicators reveal that the approach generated a large number of ideas which had been developed up to and including Stage 5 (a detailed business plan), or which, in other words, had been accepted by the organization. However, only 10 per cent of those ideas had been implemented after two years. In other words, the initial aim of dynamizing and transforming the company was not achieved.

Moreover, while pursuing a rational project, the apparatus remains fluid and reveals, during the process of its deployment, a substantial number of problems not previously taken into account (Teglborg, 2010). Thus, the amorphous nature of EDI requires that the actors involved participate in a learning process itself inscribed in a process of transformation which calls for the reflexive redesign of employee-driven innovation apparatuses.

The mirage of quantification[4]

In three-quarters of the firms studied, revenue targets were used to steer the implementation of employee-driven innovation apparatuses (Teglborg, 2010). In July 2007, during the launch of a new EDI apparatus in a large French company, representative of its sector in the same way as Solvay, with over 80,000 employees in France alone, the director made the vague suggestion that the number of ideas suggested in the old employee-driven innovation approach should be multiplied ten or 100 times. The employees reacted en masse, suggesting over 20,000 ideas in the course of a few months. Six months later, when practically none of those ideas had been implemented, the director fixed the objective of implementing 1,500 ideas by December 2007. In 2008, and then again in 2009, he demanded a number of ideas equivalent to 30 per cent of the personnel and that 15 per cent of those ideas should be implemented.

This way of operating is an example of 'the old dream of governing men as if they were things' and is characterized by a confusion between the regulation of machines and biological organisms and that of human societies underlined by Georges Canguilhem (Supiot, 2010). For Supiot (2010), number-based governance has become more important than rule-based governance. While law-based governance proposes general and abstract rules which 'guarantee the identity, freedoms and duties of all, the objective of number-based governance is the self-regulation of human societies. It is based on the faculty of *calculation*, or, in other words, on operations of *quantification* (using the same unit of calculation for different people and situations) and the *programming* of behaviours (using techniques for measuring performances: benchmarking, ranking, etc.). Under the influence of governance, normativity loses its vertical dimension: it is no longer a question of referring to a law which transcends facts, but of inferring the norm underpinning the measurement of facts' (Supiot, 2010: 78). This reduction of the diversity of people and things to a measurable quantity is, according to Supiot (2010), inherent in the project to create a total market and yet another illustration of performativity (Lyotard, 1984).

Supiot provides us with a warning: 'This attempt to metamorphose all species of singular qualities into a measurable quantity ushers us into a speculative loop in which a belief in numbered images gradually replaces contact with the realities that these images are supposed to represent' (Supiot, 2010: 81). From this perspective, identifiable objects are listed and similar methods of quantification assigned to them. In the case of a company, these identifiable objects are innovative ideas. But ideas cannot be viewed, in any sense, as mathematical entities. The operation consists of according a meaning to the process of calculation by aligning the quantities measured to a sense, or meaning, of measurement. It is thus that comparisons are made between ideas such as 'organizing aperitifs in commercial agencies' and complex technological operations involving remote control software. Unlike the use of quantification in the natural sciences, economic and social statistics do not measure a pre-existing reality but, rather, construct a new reality by treating heterogeneous beings and forces as equivalent (Desrosières, 2000).[5]

The lack of articulation between various sources of innovation

An over-emphasis on the importance of the employee as a source of innovation can lead to a lack of thought about the way in which different sources of innovation, namely clients and Research & Development departments, are articulated. Far from being a multi-source approach to innovation encompassing a strategic analysis of the way in which those sources are articulated, employee-driven innovation runs the risk of creating its own limitations and becoming a simple managerial gadget, the objective of which is to provide employees with the illusion that they are, in some manner, making a contribution.

Employee-driven innovation marks the transition from an emphasis on the results produced by innovation, in which the degree, tenor and intensity of innovations are assessed, to an emphasis on the sources of innovation (the market, technology, research, employees, etc.) (Høyrup, 2010). The accent placed on employee participation by means of workshops, systematic brainstorming sessions, challenges and prizes can, after a certain point, seem exaggerated, representing as it does a kind of fetishization of the source of innovation, namely employees. This fetishization may obscure the fact that, in spite of employee-driven innovation, companies still, strategically speaking, place an emphasis on results. Firms thus take two approaches to tempering the organizational fetishization of employees. Some call results and sources of innovation into question, while others attempt to articulate the various sources of innovation more effectively. In its deployment of employee-driven innovation, Favi applies both approaches. In 1997, the company's Managing Director introduced the project, 'How can we provide a framework for living for 100 families in Flavicourt in 2027?', thus encouraging his community of employees to invent the products of the future. The aim of the project was to generate innovations with results corresponding to the local expectations of the firm's employees. Following this initial approach, Favi promoted multi-source innovation. However, the multi-source approach turned out to be a bi-source approach, involving a combination of the desires of the firm's employees and the expectations of clients recorded by the sales force. Characterized by a certain dichotomy, this approach is very different from the pluralist approach outlined in Actor Network Theory. This is one of the risks inherent in a highly formalized approach to EDI. In a different context, Solvay-Tavaux has a substantial budget dedicated to training its employees in the field of innovation. The firm's EDI apparatus is characterized by a desire to provide the various sources of innovation – employees, local SME subcontractors, the University Institute of Technology, and experts – with a specific place in the hierarchy of processes and decisions relevant to innovation. Nevertheless, in processes of assessment of the novelty of ideas and their eventual implementation, the central concern with short-term profit once again raises its head.

In the last analysis, as we briefly mentioned above, the results of the Solvay-Tavaux experience, which appears to be the most advanced in terms of providing a variety of sources of innovation and tempering obsessions with short-term profit, were so weak that the firm took recourse to open innovation concepts in order to radicalize its multi-source, multi-channel approach to innovation. In other words, employee-driven innovation did not represent a definitive paradigm shift with regard to the way in which the firm operates.

Disillusionment and loss of interest on the part of employee–innovators

The main motivation of innovators is to see their innovations in action (Fry, 1987). But, in an organizational environment characterized by limited resources, choices must be made, a fact which implies that projects developed by employees will, of necessity, have to compete with one another (Durieux, 2000). Although they are strongly urged to suggest innovations, employees often see their ideas rejected, which may cause disappointment and an eventual loss of interest. In a large French telecommunications company, 15 per cent of the ideas suggested by employees were implemented. This was transformed into a rule (Reynaud, 2003), thus automatically condemning 85 per cent of the ideas to oblivion, regardless of their intrinsic qualities. In such a context, employees convinced of the value of their innovation projects will experience major disappointment, sometimes accompanied by suffering and anger, when their ideas are rejected by the company. It thus becomes necessary to explain, within a reasonable period of time, precisely why their ideas were not taken on board.

Furthermore, the importance of ensuring that employee–innovators feel personally fulfilled has often been emphasized (Kanter, Høyrup, 2010). However, when the possibility of innovating is transformed by the organization into an expectation, or even an obligation, EDI can lead to employees – subjected to a continual pressure to innovate – losing interest (Alter, 2000). Restrictive, systematic monitoring limits the efficiency of the approach, whereas self-monitoring encourages it. Moreover, the kind of creative learning inherent in EDI entails negative aspects. Indeed, an overemphasis on flexibility, the transformation of prevailing practices, and the creation of new solutions may create stress and feelings of anxiety and insecurity (Ellström, 2006). Moreover, Elmholdt shows that placing too much emphasis on innovative learning can have dramatic consequences if the potential for reproductive learning is neglected (Elmholdt, 2006).

Discussion

Of course, the two extreme cases used in this study partially limit our analysis in that they exemplify relatively rare phenomena. Nevertheless, both cases have the merit of revealing the chiaroscuro which characterizes EDI. Originally, the objective of employee-driven innovation was to establish a correspondence between the development of the firm and that of its employees. This correspondence presupposes the establishment of a dialogue between employees rich in propositions and realizations on a number of levels. Nevertheless, the correspondence between the development of the firm and its employees presupposes a high degree of formalization, which puts the innovation dynamic itself in peril.

Fifteen years after its introduction at Solvay-Tavaux, almost 60 per cent of the site's personnel had contributed to the EDI approach. Those participating in EDI are aware of the way in which the process has enabled them to better fulfil themselves professionally and work more effectively as part of a team. Moreover, from the point of view of innovation development strategies, EDI generates a broad range of innovative ideas. Of course, the ideas suggested are, to some extent, pre-defined; they are, after all, based on the themes chosen for challenges, which reflect the strategic objectives of the firm. But Solvay also encourages its employees to suggest situated ideas, or, in other words, ideas deriving from professional learning problems linked to specific activities and functions, and to the site itself. Lastly, Solvay welcomes ideas based on serendipity, or, in other words, intuitions and solutions associated with situations or activities which are not linked to the firm's core activities and which may, for example, involve integrating employees' hobbies into the training process.

However, it is beyond doubt that ideas reflecting strategic expectations that have been clearly signalled by the firm are more likely to be selected and accorded the kinds of resources required to implement them successfully. Furthermore, in many cases, employees are invited to readjust their ideas to take into account major organizational expectations. Perhaps the approach is a little over-hasty; evaluators do not always have the patience to incubate a large number of original, potentially successful ideas. Moreover, there is little dialogue between innovative employees and the R&D department (this was true of the development of Product 143a). Besides, it is by no means clear that the firm wants there to be. This limits the possibility of generating a greater number of radical innovations.

We thus have the impression that the purpose of ideas deriving from EDI and other innovation processes is to build up trust in the notion of innovation and even to underpin a certain myth of innovation as a means of overcoming the day-to-day problems of management. The champions of innovation at Solvay and the innovative employees at Favi are first celebrated and then either accorded roles as representatives of the company externally or made judges of challenges internally. Naturally, this is all very rewarding. Lastly, at Solvay, the management took into account the fact that 40 per cent of the firm's employees had not contributed to EDI by suggesting that a challenge concerning ideas which either had not been completed or had simply been abandoned should be organized. It should be underlined that this project focuses on recuperating good ideas that have been lost rather than on reassuring under-motivated employees. Even in cases in which attempts are made to improve the innovation process, there is always a temptation to reduce the employees' role to that of merely generating a flow of ideas.

Yet more troubling in terms of the way in which the results of EDI are presented as a managerial panacea, particularly with regard to the insistence on the notion of reducing costs, EDI appears to be a way of encouraging employees to interiorize downsizing strategies. The approach can be used to persuade employees that it is not always possible to provide them with the resources needed to attain their objectives and that they must either innovate in order to make up for this shortfall or focus less on their objectives. It is difficult to know whether this type of dynamic encourages employees to take a broader view and seek resources not supplied by the firm elsewhere, or whether, in the medium term, it is a trap which forces employees to work without being able to count on internal resources, a trap which is bound to generate a substantial amount of stress.

In spite of, or in parallel with, any temptations firms may have to use EDI as a tool of management and, secondarily, manipulation, it seems that employees and middle managers become aware of the limits and even the ambiguities of EDI: for example, 40 per cent of them displayed a standoffish attitude to Solvay's EDI apparatus. It would seem that management took the limitations of the EDI model into account. The fact that a substantial number of ideas had been implemented encouraged those responsible for managing the apparatus to envisage challenges making it possible not only to give incomplete projects a second chance, but also to ask collective questions about the logical development and ruptures characterizing the process for generating ideas. To a large degree, this approach focused more on certain problems concerning the EDI apparatus than on the difficult balance between formalization and freeing the innovative potential of employees. We would like to highlight the fact that EDI tends to obscure the risks inherent in innovation and emphasize its irenic aspects. Communication campaigns have even been known to foreground such aspects. In this context, we would insist on the fact that the processes and fruits of innovation are merely ways of reconciling employees with themselves and their colleagues and smoothing over the deleterious effects of struggles and rivalries between various departments over resources. EDI can therefore be a pretext, even a placebo, to hide organizational problems and tensions. Consequently, it is not yet possible to say whether learning is dependent on the chiaroscuro generated by the EDI process. The question to be asked, then, is what provokes this realization and, above all, in certain cases, if that realization has not come too late. In this context, the question asked by Høyrup (2010) is relevant: 'How could organizations be reconceived and redesigned in order to be optimal arenas for innovative employee behavior?' (Høyrup, 2010).

Conclusion

In this chapter, we have highlighted the effective but contrasting outcomes of EDI as well as the tensions caused by the process. The Solvay-Tavaux and Favi cases illustrate the degree to which EDI can encourage the development of competitive advantages and an environment conducive to employee learning. However, both cases suggest that EDI is a vector of inextricable tensions. The innovations suggested and developed by employees frequently conflict with organizational norms, while the predominance of the diffusion model of innovation as a managerial model (Callon, 1994) contrasts with the maelstrom-like character of innovation processes highlighted by Actor Network Theory. Similarly, the performativity of EDI apparatuses suggests that they are not entirely complete (Lyotard, 1984; Teglborg, 2010) and that they are, of necessity, part of a learning process, while the desire of managers to control the process frequently generates quantitativist problems. Lastly, despite renewed interest in EDI, its advocates have forgotten the importance of its links with other potential sources of innovations such as R&D, marketing and Open Innovation. The risk of employee–innovators becoming disillusioned and losing interest has also been underestimated.

The cases analysed largely reflect the chiaroscuro effect inherent in EDI and suggest the need for a reflective approach to EDI as a management system to the extent that reflection plays a crucial role in the learning process (Høyrup, 2006). We do not intend to suggest 'off the shelf' solutions to the task of managing EDI, but, instead, would encourage managers to familiarize themselves with the tensions inherent in the approach. EDI management apparatuses do not always produce the effects expected of them. In this sense, they can be thought of as unfaithful servants. While in no way demeaning EDI, we would encourage a reflective reconception of the use of EDI apparatuses. We thus propose reversing the process, or at least rendering it interactive by elaborating formalized representations of the practices developed in reaction to the management apparatus and presenting them to practitioners so that they can discuss them and develop new approaches. Thus, a structured comparative model will be suggested in which the real potentialities of the apparatus will be revisited and inscribed in a learning process.

Notes

1. Name given to the approach on the Tavaux site. The approach is a part of the *Innovation for Growth programme*.
2. This case is based on a presentation delivered by Jean-François Zobrist, Managing Director of Favi between 1971 and January 2009, at the Carrefour de l'Innovation Participative (EDI Crossroads) event held at Renault Boulogne Billancourt in November 2007, http://www.favi.com/.
3. Ibid.
4. Title borrowed from Supiot (2010).
5. Quoted by Supiot (2010).

References

Akrich, M., Callon, M. and Latour, B. (1988). 'A quoi tiennent le succès des innovations'. *Annales des Mines, Série Gérer et Comprendre*, 11–12.
Alter, N. (2000). *L'innovation ordinaire*. Paris: Presses Universitaires de France.
Basso, O., Fayolle, A. and Bouchard, V. (2009). 'L'orientation entrepreneuriale: Histoire de la formation d'un concept'. *Revue Française de Gestion*, 5(195).
Berry, L. (1983). *Relationship Marketing*. Chicago: American Marketing Association, 146.
Boussard, V. and Maugeri, S. (Eds) (2003). *Du politique dans les organisations*. Paris: L'Harmattan.
Broustail, J. and Fréry, F. (Eds) (1993). *Le management stratégique de l'innovation*. Dalloz.
Callon, M. (1994). 'L'innovation technologie et ses mythes'. *Annales des Mines, Série Gérer et Comprendre*, 34: 5–17.
Desrosières, A. (2000). *La politique des grands nombres. Histoire de la raison statistique*. Paris: La Découverte.
Durieux, C. (2000). *De la théorie linguistique à la théorie interprétative*. Épistémologie de la règle de trois, Cahiers de la MRSH, Presses universitaires de Caen, 11–19, 85–95
Eisenhardt, K. and Graebner, M. (2007). 'Theory Building from Cases: Opportunities and Challenges'. *Academy of Management Journal*, 50(1): 25–32.
Ellström, P.-E. (2006). 'Two logics of learning'. In Antonacopoulou, E., Jarvis, P., Andersen, V., Elkjaer, B, and Høyrup, S. (Eds) *Learning, working and living. Mapping the terrain of working life learning*. New York: Palgrave Macmillan.
Elmholdt, C. (2006). 'Innovative learning is not enough'. In Antonacopoulou, E., Jarvis, P., Andersen, V., Elkjaer, B, and Høyrup, S. (Eds) *Learning, working and living. Mapping the terrain of working life learning*. New York: Palgrave Macmillan.
Fry, A. (1987). 'The post-it note: an Intrapreneurial Success'. *SAM Advanced Management Journal*, 52(3).
Hatchuel, A. and Weil, B. (1994). 'L'expert et le système, suivi de quatre histoires de systèmes-experts'. *Revue française de sociologie*, 35(35–1): 137–139.
Høyrup, S. (2006). 'Reflection in learning at work'. In Antonacopoulou, E., Jarvis, P., Andersen, V., Elkjaer, B, and Høyrup, S. (Eds) *Learning, working and living. Mapping the terrain of working life learning*. New York: Palgrave Macmillan.
Høyrup, S. (2010). 'Employee driven innovation and workplace learning: basic concepts, approaches and themes'. *Transfer*, 16(2): 143–154.
Latour, B. (1991). *Nous n'avons jamais été modernes. Essai d'anthropologie symétrique*. Paris: La Découverte.
Letiche, H. (2009). 'The dark side of organizational knowing'. E: CO, *Forthcoming*, 11(3), 59–70.
LO (2007). *Employee-driven-innovation-a trade union priority for growth and job creation in a globalised economy*. Copenhagen: LO publishing.
Lyotard, J.-F. (1984). *The Postmodern Condition: A Report on Knowledge*. Bennington, G. and Massumi, B. (trans.). Minneapolis: University of Minnesota Press.
Moisdon, J.-C. (1997). *Du mode d'existence des outils de gestion*. Paris: Seli Arslam.
Moisdon, J.-C. (2003). *Sur la largeur des mailles du filet: savoirs incomplets et gouvernement des organisations*. Gouvernement, organisation et gestion: l'héritage de Michel Foucault. Laval: Les Presses de l'Université Laval.
Nieuwenhuis, L. (2002). Innovation and learning in agriculture. *Journal of European Industrial training*, 26(6): 283–291.

Reynaud, E. (2003). *Quand l'environnement devient stratégique.* Session semi-plénière, Congrès de l'Association Internationale de Management Stratégique, June, Université de Tunis.
Serres, M. and Latour, B. (1995) *Conversations on Science, Culture, and Time.* The University of Michigan Press.
Siggelkow, N. (2007). Persuasion with case studies. *Academy of Management Journal,* 50(1): 25–32.
Supiot, A. (2010). *L'esprit de Philadelphie-la justice sociale face au marché total.* Paris: Seuil.
Teglborg, A.-C. (2010). *Les dispositifs d'innovation participatives. Une reconception réflexive à l'usage.* Ecole doctorale Paris I – IAE de Paris – HEC Paris, Paris.
Vaughn, B.E. (1999). 'Power is knowledge (and vice versa): A commentary on "Winning some and losing some": A social relations approach to social dominance in toddlers'. Merrill- Palmer Quarterly, 45, 215–225.
Yin, R.K. (1989). *Case study research: Design and methods* (rev. edn). Beverly Hills, California: Sage Publishing.

3

In Search of Best Practices for Employee-Driven Innovation: Experiences from Norwegian Work Life

Tone Merethe Aasen, Oscar Amundsen, Leif Jarle Gressgård and Kåre Hansen

In this chapter we identify and discuss important elements of Employee Driven Innovation (EDI), and show how the interrelationship between professional role performance, cultural characteristics, and supportive means and tools is central in this respect. Drawing on data from 20 Norwegian enterprises known for their productive involvement of employees in innovation work, we discuss how leaders, employees and union representatives can carry out their work and use various means and tools to encourage the development of cultural characteristics essential for successful EDI practices. Our study indicates that EDI is mainly about how managers and employees see and perform their roles, and less about formal structures. There are several organizational efforts that can be made to support the development of a culture for joint innovation effort, but improved innovation capacity through the implementation of EDI practices requires the successful interplay between all three dimensions (roles, culture and tools).

Introduction

The potential of enterprises to produce innovative, commercially or otherwise valuable results is referred to as their capacity to innovate (Neely *et al.*, 2001). This capacity has been defined as a multidimensional, complex variable, which may be connected, among other things, to the resources and capabilities of an enterprise, and its ability to use these to explore and exploit opportunities (Grant, 1991; Prahalad and Hamel, 1990; Verona and Ravasi, 2003). Although this is an indication of the essential role of individuals in knowledge development processes, the literature tends not to focus on the importance of employee participation in innovation processes, or the premises for such participation (Byrne *et al.*, 2009).

The Norwegian Confederation of Trade Unions defines employee-driven innovation (EDI) as innovations (new products, processes or services) brought about through an open and including innovation process, based on a systematic use of ideas, knowledge and experience of employees, which have a positive impact on the organization's overall innovation capability. Inherent in this definition is the fundamental assumption that employees have competencies and ideas that may strengthen an organization's capacity to innovate, given that the conditions are favourable. This assumption is deeply rooted in a long tradition in Scandinavian work life, where involvement of employees in development activities has been central. In spite of this, EDI seems to be an underexplored opportunity in many organizations, in both private and public sectors.

In this paper we investigate and describe successful practices for EDI. Drawing on qualitative interviews with employees and leaders from 20 Norwegian enterprises, we seek to draw more general conclusions about characteristics of organizations that have succeeded with EDI practices. The question guiding our research has been: *How can successful practices for EDI be described?* The paper has four parts: (1) a theoretical section presenting some of the existing literature within innovation and EDI research relevant to this study; (2) a methodological section, covering the approach to research; (3) an analytical section, in which the results are explained by means of three dimensions (professional roles, cultural characteristics and tools supportive of EDI practices); and (4) a concluding section, pinpointing some lessons from the study.

Theory

The dominant view of innovation is that it is a development process initiated on the basis of a new idea and terminated by the introduction in a market of a material or immaterial invention (Goffin and Mitchell, 2005; Kelley and Littman, 2005; Tidd and Bessant, 2009). In general, the ability to innovate is regarded as imperative for the competitive advantage and sustainability of organizations in both public and private sectors, and innovation has therefore been a subject of major research efforts during the last decades. There are generally two main models of innovation studies; normative and descriptive. Normative models are prescriptive in nature, and usually provide guidelines for the design and development of organizations to increase innovative capacity (Drucker, 2002; Kanter, 1983; 1988; Quinn, 1985). The descriptive models, on the other hand, summarize the characteristics observed regarding innovative organizations and processes for innovation, and relationships between such characteristics (Dougherty and Hardy, 1996; Neely *et al.*, 2000; Van de Ven *et al.*, 1989; von Hippel, 1988). As described by Rothwell (1994), the models suggested for innovation processes have changed with changes in society or in the context of innovation. Today, invention and innovation are increasingly understood as a result of the exchange of knowledge between different actors within an organization, and in different organizations (Caloghirou et al., 2004; Hargadon, 2003; Powell, 1998). Most descriptive or empirical models of innovation, including non-linear models, identify the following innovation phases: *idea phase, development phase* and *implementation phase*. In addition, recent models usually include an *experience phase* to emphasize the importance that learning is made explicit and exploited as the basis for improvement work and further innovation.

While research about external sources of innovation, such as user-driven innovation (Von Hippel, 1988; 2005) and open innovation (Chesbrough, 2003), has attracted much attention, EDI is a relatively new field of research and the number of contributions is limited. Current research on employee involvement in innovation can be divided into two main strands. The first is about conditions for EDI, internally and related to the organizational context (Byrne et al., 2009; De Jong and Kemp, 2003; Pelz, 1956; Smith et al., 2008; Tierney et al., 1999). The second strand focuses on consequences of EDI (Hamel and Prahalad, 1994; Heller et al., 1998; Freeman and Soete, 2000; Kelley, 2010). Hardly any of the contributions discuss the potential effects of EDI on innovation capacity or value creation. This could be explained by the complexity of such research, involving studies of a comprehensive set of interdependent and independent factors, affecting the outcome of innovation work in various ways. However, there are several studies, including work research, suggesting that internal factors such as autonomy, collaboration, organizational climate, and so on have positive effects on individual motivation and satisfaction (Axtell et al., 2000; Smith et al., 2008; Woodman et al., 1993), and on organizational factors such as turnover, quality of work and sickness absence (Black and Lynch, 2001; Meyer and Allen, 1997; Østberg et al., 2010; Tidd and Bessant, 2009). Concerning capacity for innovation, some studies indicate that competition and market conditions will affect the outcome of EDI processes (Acs and Audretsch, 1987; De Brentani, 2001; De Jong and Kemp, 2003).

Methodological approach

The objective of the present study was to describe and analyse important premises for the successful involvement of employees in innovation processes. Implicitly, the conception of 'best practice' in the study is closely related to the process of case selection. The selection was not based on the objective of ensuring a representative sample of Norwegian organizations, but rather on what is referred to as *strategic* or *purposeful* selection (Flyvbjerg, 2006; Morrow, 2005). In terms of Flyvbjerg (2006: 230), we did an 'information oriented selection', meaning that cases were selected on the basis of expectations about their (rich) information content. In addition to the research team's broad knowledge about Norwegian enterprises, a reference group actively participated in the identification of relevant cases. The reference group consisted of representatives from the Norwegian Ministry of Trade and Industry, The Norwegian Confederation of Trade Unions, and The Confederation of Norwegian Enterprise.

The final decision about which cases to include in the study was made by the research team. A total sample of 20 enterprises was selected, representing both public and private sectors; different sizes (number of employees and annual sales); different industries; and located in different geographical areas in Norway. Furthermore, enterprises with and without union representatives were included. The choice of several heterogeneous cases was made to increase the possible generalizability of results across cases (Schofield, 2002). Table 3.1 provides an overview of the cases, including number and type of informants.

An interview guide was developed on the basis of a theoretical framework, which included innovation process phases and factors previously identified as important for EDI (De Jong and Kemp, 2003: Smith *et al.*, 2008; Tierney *et al.*, 1999). With the exception of two of the interviews, semi-structured group interviews with two or three respondents were carried out. The decision to use group interviews was based on two considerations. First, the topic was not sensitive in a way that would require individual interviews. The second consideration was linked to the assumed advantage of this approach, involving a 'naturalistic' setting where the informants could build on each other's experiences during the interview (Gaskell, 2000).

The duration of each interview was 1 to 1.5 hours. As shown in Table 3.1, a total of 48 informants were interviewed, and each interview was conducted by two researchers. The informants were leaders, employees and union representatives. In organizations with union representatives, the three groups were represented together in 60 per cent of the interviews. All interviews were recorded and subsequently transcribed. Data were coded based on a pre-defined scheme and analysed with reference to innovation phases and the thematic categories selected in the theoretical framework. The two researchers conducting each interview were jointly responsible for the preliminary content analysis, and the results were subsequently shared and discussed between all four researchers.

Morrow (2005) underlines that the validity of qualitative inquiry is related to the information richness of the cases selected, as well as to the analytical capabilities of the researcher(s). The analytic phase in the present study was improved by the development of work processes where the multidisciplinary team of four researchers worked closely together, alternating between individual analysis and joint discussions based on empirical data and a broad theoretical understanding. This way of approaching research has been referred to as 'mutual construction of meaning between co-researchers' (Rismark and Sølvberg, 2007: 602).

Table 3.1 Case enterprises

Industry (NACE classification)	Company	Sector Private	Sector Public	Size (employees)	Informants Leader	Informants Employee	Informants Union rep.
Administrative and support service activities	1		X	50–100	1	2	
Construction	2	X		>250	1	1	1
	3	X		>250	1	1	
	4	X		50–100	1	1	
	5	X		50–100	2		
Electricity, gas, steam and air condition supply	6		X	>250	1	1	
Financial and insurance activities	7	X		10–50	1	1	
Information and communication	8	X		>250	1	1	
	9	X		100–250	1	2	
	10	X		50–100	1	2	
Manufacturing	11	X		>250	1	1	1
	12	X		10–50	1	2	
	13	X		10–50	1	2	
	14	X		>250	1	1	1
	15	X		>250	1	2	1
Professional, scientific and technical activities	16	X		50–100	1	1	
	17	X		50–100	1	1	1
	18	X		>250	2	2	
Public administration and defence; compulsory social security	19		X	>250		1	
	20		X	>250	1	2	

Results and discussion

In the data analysis, a number of elements appeared as particularly important for successful innovation work in general, and EDI in particular. These elements can be categorized into three interrelated domains of organizational conduct, which structure the following discussion. The first domain concerns the performance of three types of professional *roles* in organizations: leaders, employees and union representatives. Second, we found that, among our case organizations, certain *cultural characteristics* could be recognized, although not *all* characteristics were present in all organizations. In this context, the concept of organizational culture is understood in general terms, as shared networks of meaning or basic assumptions guiding people into certain patterns of thought. In other words, culture is seen as traditions of acting and thinking in particular ways (Alvesson and Sveningsson, 2008; Jaffee, 2001). The empirical background for using culture as a concept in the analysis is related to the rather *wide range of specific practices* found in the cases (i.e. how EDI was carried out in the different enterprises). To extract something meaningful and possibly general about the realization of EDI in this context, the concept of culture was considered as a possible and even promising approach. However, the basis for this choice was also found in the discourse of the cases; informants from different enterprises used the term 'culture' in the interviews to characterize their realization of EDI, indirectly suggesting that the idea and concept should be considered more closely in the research group. The third domain concerns the application of specific tools to encourage and facilitate EDI practices. Such tools may constitute the backbone of innovation work, given a productive interplay between cultural characteristics, performance of roles, and the selected tools. In this context, 'tools' spans several kinds of measures for information exchange, idea registration and evaluation, and improved collaboration.

Roles in the organization

Roles may be defined as a set of expectations directed towards a certain position (Goffman, 1959), and performing a role is thus the manifest way a person acts in his or her position. Our results indicate that the most important leader characteristics in organizations successfully involving employees in innovation work are openness towards change and confidence in the delegation of responsibility. Allowing the employees freedom and opportunity to develop and follow-up ideas, and letting processes be carried out without excessive control and governance, appear to be fundamental in this respect:

> I do not participate in the process. I know that if I participate, then I will be managing the process. And then I'm not going to get the necessary

enthusiasm and involvement needed, so therefore, I prefer to be a conversation partner. (Manager, Company 14)

Involvement of employees in innovation work necessitates an open and constructive dialogue between employees and leaders (Tierney et al., 1999), which means that the role of the leader must be more cooperative than instructive. This is confirmed in many of our interviews and illustrated by the following statement from one of the managers:

> One needs to have an excellent ability to lead conversations and to be caring. To be direct and firm, yet have the ability to listen to the employees. And to be a person who is involved, also in the individual's well-being. (Manager, Company 3)

Furthermore, employee involvement implies a certain delegation of authority, meaning that there are processes in the organization the leader should not engage in. As a consequence, the leader has to clearly communicate his or her expectations about contributions (i.e. suggestions for improvements and cooperation) from the employees, ensure that contributions are given credit, and make the results visible in the organization:

> What you don't want to do as a manager when your employees have started to become more motivated, is to finish the final 20 per cent of the work after everyone has gone home. Then you've made a big mistake. It is the most difficult tasks that remain. What's fun is done. So a good advice is to let them take everything. Don't clean up and finish the job. (Manager, Company 20)

Our data further indicate that the role of working management is of particular importance for EDI. The working manager (i.e. foreman, group leader, middle manager, etc.) is the one closest to the daily operations, and therefore the one having everyday responsibility for capturing ideas and suggestions for improvement. The working manager is also decisive when it comes to inspiring and motivating the employees to be engaged and show initiative at work. The most important challenges for the working managers are to extend their knowledge of the business outside their specialist area, and to master the demanding combination of performing their job as a skilled worker and a capable leader at the same time.

Our results also indicate that successful EDI is related to how employees see and perform their roles. Employees' consciousness of their importance to the performance of the firm is central in this respect, which highlights the concepts of responsibility and involvement. Further, we also find that employees' sense of responsibility and belonging is strengthened when they are offered extended insight into their organizations, including financial and strategic aspects.

> I feel that we really don't have a hierarchy in our company. We are all members of the same company, and I think everybody feels a great responsibility for their jobs, and that we all have an equal opportunity to contribute. Everyone realizes that it is us who create the company, it's not just our boss. (Employee, Company 10)

As the citation suggests, employees' willingness to take responsibility beyond their main tasks is related to their consciousness of the importance of their contribution to the development of an interesting and long-term place to work. An emphasis on employees' responsibilities, in terms of their own work but also for seeing their role in a wider context, is therefore important for EDI. Involvement implies not only that ideas and contributions from employees are demanded, but also that employees are given responsibility for creating arenas where contributions from people external to the organization can be captured and communicated in the organization.

Another important professional group is the union representatives. First, by being the management's discussion partners, they can contribute to a well-functioning cooperative relationship between their members and the management, being fruitful for all parties. A constructive relationship between the union representatives and the management depends on the mutual understanding that the purpose of involvement and cooperation is the securing of long-lasting, good working conditions:

> The fact that we as union representatives always wanted to contribute, and to be part of the strategy and planning processes, has been a driving force in this company. (Union representative, Company 17)

Moreover, engaging in dialogue with the union's members in order to inform and prepare for forthcoming events and processes in the organization is likely to increase the employees' 'organizational overview' (i.e. understanding of the performance of the firm, as well as seeing their role in a broader context).

Cultural characteristics

When analysing the interview transcripts we found that organizations that have successfully adopted EDI practices have a number of cultural characteristics in common. These characteristics are interrelated, and can be hard to distinguish. Thus, changing the nature of one of them may result in changes to the others. We have labelled the first three cultural features *commitment, cooperativeness* and *pride*. In the organizations we were studying, we found that employees were willing to make an extra effort in order to improve, and to reach organizational goals. In other words, they had a sense of belonging to the workplace. There was also a general understanding among the informants that a cooperative environment in terms of constructive dialogue and involvement (across horizontal and vertical borders in the organization) was essential for obtaining the best results. Employees and leaders clearly expressed that they were proud of their workplace. They also acknowledged the importance of the organization's reputation and the implicit need for them to be good ambassadors, which again had positive effects on the perceived importance of their work.

Trustfulness, autonomy, tolerance and *feeling of security* were found to be another set of important cultural features. Employees, leaders and union representatives all pointed out the mutual interdependency of the various professional roles in the organization, and emphasized the importance of building trustful relations. The concept of responsibility was important in this respect. That is, leaders were delegating decision-making powers to the employees, and trusted their ability to carry out the work, and this was accepted and appreciated by the employees. This distribution of responsibility and work group autonomy was found to release a considerable amount of creativity and energy among employees. Furthermore, in successful EDI organizations, diversity was seen as an advantage, and it was acknowledged that people are different in many respects and thereby also have complementary skills. The organizations were further characterized by a tolerance for failure, a climate for information-sharing, and free expression of ideas, which again was seen as important for organizational learning and ability for improvement. The citation below illustrates the importance of a supportive climate and a feeling of safety and security:

> There is no culture for giving up. In this company we stand together and keep on working. Of course you experience setbacks from time to time when working in projects. But then people immediately back you up, encourage you to keep going on, and tell you that this is but a valuable lesson. This kind of support and tolerance for failure is necessary for companies to gain long-lasting advantages and success with projects. Quitting or giving up is an option, but I don't think that we have ever done that. (Employee, Company 8)

The citation above is also related to another cultural feature, which is *learning orientation*. Implicitly, learning and improvement are seen as integral parts of work. We found this to be related to the perception of change in an organization; that is, whether modification and changes at the workplace represented an opportunity or a threat. Several leaders emphasized the importance of building a culture where change was perceived as positive and representing an opportunity. However, clearly communicating that 'change does not mean that you will lose your job' was seen to be of vital importance in this respect.

The final cultural feature identified was *openness*. Our informants clearly related the idea of 'openness' to the successful implementation of EDI practices in their organizations, but the concept was interpreted in different ways. As an example, it was related to widespread involvement and transparency in the organization concerning decision-making processes. Other examples were the establishment of intranet discussion forums, where employees were given the opportunity to discuss organizational affairs very openly, and the employees' experience that their leader's door was always open if they wanted to discuss an idea.

Means and tools (EDI toolbox)

The case companies applied a variety of means and tools to involve employees in innovation work, and to support the development processes. The means and tools could be categorized into four groups: tools for idea registration and development; tools for idea prioritizing; tools for dissemination of information; and tools supporting an innovation-focused culture. Moreover, tools were of three main types:

- Manual tools, such as lists and forms, newsletters or message boards.
- E-tools, such as dedicated computer software, social media, tailored forms based, for example, on Microsoft Word/Excel, intranet or information screens.
- Tailored processes and systems, such as designation of idea brokers, adapted education, research projects, formal and informal routines for idea development and invention, idea and concept development methodology, and HR policies.

In addition to these tools, we identified formal and informal meetings as important arenas for the development and implementation of good practices for employee-driven innovation.

The tools most commonly used were tools for idea registration and development. These were tools associated mainly with the first phase of innovation, and usually seen to be about ideas for new or improved products or work processes. However, ideas were also important to solve all the small and bigger problems emerging during innovation work. The most important management arguments for the implementation of tools to register ideas were that it made the ideas visible to the organization, and that it represented a way not to lose ideas that were not followed up immediately.

E-tools open to all employees were seen as particularly suited for joint idea development. Most of the tools in this category were used by employees only, but some companies registered ideas originating from customers, partners or suppliers, and some even opened their systems for externals to register ideas themselves. For such tools to have the intended effect, two factors appeared to be of particular importance. One was to minimize the barriers of use, related, for instance, to the fact that not all employees use a PC as part of their everyday work. In addition, experience was that some employees did not feel comfortable when expected to formulate ideas in writing, especially if they had not mastered the native language. The second factor was the genuine and continued management engagement, encouraging employee involvement and empowering employees to spend time on idea exploration.

Another characteristic of organizations having implemented EDI practices in a successful way was a concurrent focus on idea capturing and idea prioritizing. Often, idea prioritizing involved decisions about improvements of work processes and technology, and, more often than not, employees carried out such changes without the involvement of managers. On the other hand, if idea realization implied a need for more comprehensive use of resources, management participation in decision processes was essential. Important conditions for the successful use of idea prioritizing tools were the active communication with the originators of ideas, even if the idea was not pursued, and the handling of potential issues concerning intellectual property (IP).

The advantageous use of EDI practices was closely associated with employees' access to information about all aspects of the company. In the most successful organizations, everyone was expected to look beyond their own tasks and expertise, and consider their possible contributions to the intention and future development of the organization. As with the use of any EDI tool, managers' prioritized focus on communication stood out as essential. One aspect of this, however, was the potential situation of information overload. This made the issue of dissemination of information an important question about information need, and also about who should define such need.

The preceding tool categories may support the development and continuation of a suitable EDI culture. However, these were commonly combined with means and tools encouraging the general focus on innovation work, independent of specific problems or development processes. For such tools to lead to the intended effects, leaders' engagement is not sufficient. We found that employees' perception of responsibility for the organization's achievements and profitability is equally important. Furthermore, suitable routines for use and follow-up of means and tools should exist.

Relationship between roles, cultural characteristics and tools

As discussed, our study indicates that the fundamentals of EDI could be defined by the three interrelated elements of *roles*, *tools* and *culture*. This means that the performance of professional roles and the use of adapted tools may influence cultural characteristics supportive of successful EDI practices. A major point in this is that organizational culture cannot be changed through 'resolutions'. What everyone *can* influence is the way they perform their roles; leaders can, for example, choose the adoption of tools to support the change work. Importantly, the development of EDI practices will thus concern the way professional roles should be performed, and, as organizational members change their approaches to work, this will influence the need for tools and further impact on organizational culture. Figure 3.1 illustrates the interrelationship between the elements of EDI.

Whether used independently or in combination, EDI tools were perceived to support employee involvement in innovation, and in particular the increased engagement of employees in idea communication and exploitation. Yet, it should be emphasized that the adoption of such tools will not ensure good EDI practices in itself. It is the successful interplay between management and employees' practices, organizational culture and suitable tools which leads to improved innovation capacity. We see some of the tools as supportive aids in the process of developing an organizational culture facilitating the involvement of employees in innovation. Other tools will be suitable when the organization has reached a more advanced level of EDI, to ensure the continued focus on involvement.

A particular management challenge related to the implementation of EDI tools and practices is that increased employee involvement and empowerment do not necessarily lead to increased profitability in the short term. Quite a few of our respondents told about relatively long periods of transition characterized by employees' frustration, insecurity and anger (and even by declining profits), before curves started to turn upwards. As a consequence, a fundamental condition for succeeding with the implementation of EDI practices is that managers, union representatives (when relevant) and (a sufficient number of) employees share the belief that the change they

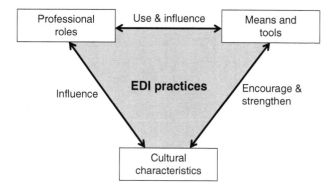

Figure 3.1 Interrelated elements of EDI

are about to accomplish is the right thing to do. Among important tools that may facilitate this condition are managers' delegation of responsibility, collaboration across departments and tasks, employees' freedom to act, appreciation of enthusiasts/enthusiasm, and positive profiling of important results (even the small ones).

Although our study does not give any indication that there are specific tools which are generally more suitable than others for the successful implementation of EDI, we found that successful EDI-practising companies were able to combine the use of tools in complementary ways. We further found that processes sometimes were accomplished without the support of formal tools, reflecting an embedding of EDI practices in the culture rendering some kinds of tools, for instance for supporting the generation of ideas, superfluous. It should also be emphasized that the most important of all tools, probably, is conversation. A constructive conversation in this respect is characterized by a mutual will to impart knowledge, and to listen. The quotation below is one of many examples offered by the informants of the importance of conversation:

> If you work on something, you don't always think about it as an idea. But then you talk with your boss about it, or maybe with someone else. And then they realize that this is an idea. (Employee, Company 10)

An underlying and highly relevant theme in the discussion of the relationship between culture, roles and tools is related to the question of intentional cultural change happening in general. Consistent with the understanding of 'culture' adopted in this article, cultural change can be viewed as 'reframing of everyday life' in Alvesson's (2002) sense. The most significant consequence of this point of view, in our context, is that culture is changed through direct interactions between people in everyday working life. This implies that a leader or another person working for change mainly has an impact on the persons they interact with directly.

This has consequences for both the use or selection of tools and the understanding of roles in the context of supporting cultural change in the organization. Dealing specifically with the question of cultural change, Alvesson and Sveningsson (2008) point out some interesting advice. Their first reminder is to insist that everybody is included in the change and 'not just those to be "worked" upon for improvement' (ibid.: 176). The point is that leaders, employees and union representatives should all view themselves as part of the change process, and to avoid an understanding which implies that the cultural change process is about an enlightened elite getting 'the other' (the rest, or the organization) to change.

Another aspect is related to the acknowledgement of culture as a slow-moving phenomenon, and therefore the need for a long-term view of the change work (ibid.). This relates especially to the role of the leader in keeping cultural themes 'permanently' on the agenda. The point here is not a demand to use a lot of energy on the cultural change work every day, but not to 'tick it off' as a completed task after the first effort. Alvesson and Sveningsson underline the need for efforts of 'listening' among leaders in an organization working on cultural change. Since change processes in this field can hardly be managed through strict and pre-defined plans, there will be a need to review the process carefully, and to revise plans on the basis of input from responses and how the ongoing work is being 'read' among employees. The role of both the union representatives and the working managers as 'listening posts' can be central for the management to understand how the process is developing. The concrete tools to support this work can be forums and meeting-places characterized by open agendas. This also relates to the crucial achievement of establishing a strong sense of 'we' to succeed in cultural change work (ibid.). Selecting core participants in the projects and work on cultural change might be a critical factor when it comes to anchoring the process broadly in the organization. Alvesson and Sveningsson warn about a practice in which people who might be conceived as 'peripheral' in the organization are the only ones being engaged directly in the cultural work process (consultants, HR people etc.). This might lead to cynical responses and questions about relevance and credibility for the change initiative among employees. This implies using tools that include both 'typical employees' and working managers in the work process, that is, thinking participation in both an indirect and a *direct* sense (Wilkinson and Dundon, 2010).

Concluding remarks

The study presented in this chapter has been a search for best practices for employee-driven innovation. Analysis of data from 20 companies successful in involving employees in innovation work shows that they have some characteristics in common concerning their role performances, their traditions of acting and thinking in particular ways, and the structural mechanisms (tools) they use to involve employees in innovation processes.

At the same time, our study clearly indicates that there is no single best practice for EDI. The companies we have investigated share some important characteristics, but they also differ in several substantial ways. Some of them have a highly structural approach to EDI, while others are characterized by a lack of tools but a strong cultural dynamic. Still, we find that our case studies represent important sources of knowledge, not in terms of fixed rules, but in the ways they give insight into the interrelationship between roles, cultural aspects and tools as the main components of EDI. In the introduction of the paper, the main objective was stated to be drawing general conclusions about characteristics of enterprises that have succeeded with EDI practices. In light of the analysis and discussion of results, we conclude that characteristic of enterprises profiting from fruitful EDI practices is the successful integration of the elements of culture and role performance, combined with use of appropriate tools in their innovation work. In other words, focusing on the complementarity of these elements is fundamental to promoting EDI in organizations.

Given the fact that our study was conducted in Norway, with its distinct traditions of employee involvement, there are obvious reasons to end this chapter with some remarks regarding the possible limitations for generalization of results to other countries with different models for employee participation. One could easily state that the Scandinavian model of cooperation and participation in working life represents a convenient framework for the development of EDI. A number of the same factors that have a positive impact on EDI are identical to some of the characteristics associated with the Scandinavian model; high-trust organizations, broad involvement of employees and a cooperative climate between management and union representatives. However, we have no reason to believe that this overlap makes our findings irrelevant to organizations in other countries. Our general suggestion is that organizations interested in adopting good EDI practices should start by assessing their present situation, with a focus on established meeting arenas, collaborative climate, and so on. The next step would then be to evaluate existing means and tools, and determine which of these are seen as most suited to support the desired development. We also recommend companies to consider organizational moves that can be made to support the development of good practices, and a culture for joint innovation effort. Examples of such steps have been presented earlier in this chapter.

Finally, a comment should be made on the need for further research. As indicated, research on EDI is still in an early phase, and more studies combining qualitative and quantitative approaches are needed. First, there is a lack of qualitative knowledge about organizational conditions promoting EDI practices. The development of this kind of knowledge is resource-demanding, because it requires that researchers participate in organizational everyday life for longer periods of time to observe how leaders and employees cooperate for innovation. Such insight will form the basis for new knowledge about mechanisms decisive for the success of EDI. Second, documentation of quantifiable profit of EDI practices, on both an organizational and a social level, is needed. This involves a particular challenge concerning the need to identify and separate effects of EDI from other factors influencing the outcome of innovation efforts in organizations, but should nevertheless be striven for.

References

Acs, Z.J. and Audretsch, D.B. (1987). 'Innovation, market structure, and firm size'. *The Review of Economics and Statistics*, 69(4), 567–574.

Alvesson, M. (2002). *Understanding Organizational Culture*. London: Sage Publications.
Alvesson, M. and Sveningsson, S. (2008). *Changing Organizational Culture. Cultural Change Work in Progress*. London: Routledge.
Axtell, C.M., Holman, D.J., Unsworth, K.L., Wall, T.D., Waterson, P.E. and Harrington, E. (2000). 'Shopfloor innovation: facilitating the suggestion and implementation of ideas'. *Journal of Occupational and Organizational Psychology*, 73, 265–285.
Black, S. and Lynch, L. (2001). 'How to compete: the impact of workplace practices and information technology on productivity'. *Review of Economics and Statistics*, 83(3), 434–445.
Byrne, C.L., Mumford, M.D., Barrett, J.D. and Vessey, W.B. (2009). 'Examining the leaders of creative efforts: what do they do, and what do they think about?' *Creativity and Innovation Management*, 18(4), 256–268.
Caloghirou, Y., Kastelli, I. and Tsakanikas, A. (2004). 'Internal capabilities and external knowledge sources: complements or substitutes for innovative performance?' *Technovation*, 24, 29–39.
Chesbrough, H. (2003). *Open Innovation. The New Imperative for Creating and Profiting from Technology*. Boston, Massachusetts: Harvard Business School Press.
De Brentani, U. (2001). 'Innovative versus incremental new business services: different keys for achieving success'. *Journal of Product Innovation Management*, 18(3), 169–187.
De Jong, J.P.J. and Kemp, R. (2003). 'Determinants of co-workers' innovative behaviour: an investigation into knowledge intensive services'. *International Journal of Innovation Management*, 7(2), 189–212.
Dougherty, D. and Hardy, C. (1996). 'Sustained product innovation in large mature organizations: overcoming innovation-to-organization problems'. *Academy of Management Journal*, 39(5), 1120–1153.
Drucker, P.F. (2002 [1985]). 'The discipline of innovation'. *Harvard Business Review*, August, 95–102.
Flyvbjerg, B. (2006). 'Five misunderstandings about case-study research'. *Qualitative Inquiry*, 12(2), 219–145.
Freeman, C. and Soete, L. (2000). *The Economics of Industrial Innovation*. 3rd edn. Boston, Massachusetts: MIT Press.
Gaskell, G. (2000). 'Individual and group interviewing'. In Bauer, M.W. and Gaskell, G. (Eds) *Qualitative Researching – With Text, Image and Sound*. London: Sage Publications.
Goffin, K. and Mitchell, R. (2005). *Innovation Management: Strategy and Implementation Using the Pentathlon Framework*. New York: Palgrave Macmillan.
Goffman, E. (1959). *The Presentation of Self in Everyday Life*. New York: Anchor Books.
Grant, R.M. (1991). 'The resource-based theory of competitive advantage: implications for strategy formulation'. *California Management Review*, 33, 114–135.
Hamel, G. and Prahalad, C.K. (1994). *Competing for the Future*. Boston: Harvard Business School Press.
Hargadon, A. (2003). *How Breakthroughs Happen. The Surprising Truth about How Companies Innovate*. Boston, Massachusetts: Harvard Business School Press.
Heller, F., Pusic, E., Strauss, G. and Wilpert, B. (1998). *Organizational Participation: Myth and Reality*. Oxford: Oxford University Press.
Jaffee, D. (2001). *Organization Theory. Tension and Change*. Boston, Massachusetts: McGraw Hill.

Kanter, R.M. (1983). *The Change Masters*. Boston, Massachusetts: Unwin Paperbacks.
Kanter, R.M. (1988). 'When a thousand flowers bloom: structural, collective, and social conditions for innovation in organization'. *Research in Organizational Behavior*, 10, 169–211.
Kelley, B. (2010). *Stoking Your Innovation Bonfire*. Chichester: Wiley.
Kelley, T. with Littman, J. (2005). *The Ten Faces of Innovation: IDEO's Strategies for Defeating the Devil's Advocate and Driving Creativity Throughout Your Organization*. New York: Currency.
Meyer, J.W. and Allen, N.J. (1997). *Commitment in the Workplace: Theory, Research and Application*. Thousand Oaks, California: Sage Publications.
Morrow, S.L. (2005). 'Quality and trustworthiness in qualitative research in counseling psychology'. *Journal of Counseling Psychology*, 52(2), 250–260.
Neely, A., Fillipini, R., Forza, C., Vinelli, A. and Hii, J. (2001). 'A framework for analyzing business performance, firm innovation and related contextual factors. Perceptions of managers and policy makers in two European regions'. *Integrated Manufacturing Systems*, 12, 114–124.
Østberg, L., Robinson, A.G. and Schroeder, D.M. (2010). *Små ideer – stora resultat*. SIS Förlag.
Pelz, D.C. (1956). 'Some social factors related to performance in research organizations'. *Administrative Science Quarterly*, 1, 310–325.
Powell, W.W. (1998). 'Learning from collaboration: knowledge and networks in the biotechnology and pharmaceutical industries'. *California Management Review*, 40(3), 228–240.
Prahalad, C.K. and Hamel, G. (1990). 'The core competence of the corporation'. *Harvard Business Review*, 68, 79–93.
Quinn, J.B. (1985). 'Managing innovation: controlled chaos'. *Harvard Business Review*, 63, 73–84.
Rismark, M. and Sølvberg, A.M. (2007). 'Effective dialogues in driver education'. *Accident Analysis and Prevention* 39, 600–605.
Rothwell, R. (1994). 'Towards the fifth-generation innovation process'. *International Marketing Review*, 11(1), 7–31.
Schofield, J.W. (2002). 'Increasing the generalizability of qualitative research'. In Huberman, A.M. and Miles, M.B. (Eds) *The Qualitative Researcher's Companion*. Thousand Oaks, California: Sage Publications.
Smith, P., Kesting, P. and Ulhøi, J.P. (2008). 'What are the driving forces of employee-driven innovation?' *Presented at the 9th International CINet Conference*, Valencia, Spain, 5–9 September.
Tidd, J. and Bessant, J. (2009). *Managing Innovation. Integrating Technological, Market, and Organizational Change*. 4th edn. Chichester: Wiley.
Tierney, P., Farmer, S.M. and Graen, G.B. (1999). 'An examination of leadership and employee creativity: the relevance of traits and relationships'. *Personnel Psychology*, 52, 591–620.
Van de Ven, A.H., Angle, H.L. and Poole, M.S. (Eds) (2000[1989]). *Research on the Management of Innovation. The Minnesota Studies*. 2nd edn. New York: Oxford University Press.
Verona, G. and Ravasi, D. (2003). Unbounding dynamic capabilities: an exploratory study of continuous product innovation. *Industrial and Corporate Change*, 12, 577–606.
von Hippel, E. (1988). *The Sources of Innovation*. New York: Oxford University Press.

von Hippel, E. (2005). *Democratizing Innovation*. Boston: MIT Press.
Wilkinson, A. and Dundon, T. (2010). 'Direct employee participation'. In Wilkinson, A., Gollan, P.J., Marchington, M. and Lewin, D. (Eds) *The Oxford Handbook of Participation in Organizations*. New York: Oxford University Press.
Woodman, R.W., Sawyer, J.E. and Griffin, R.W. (1993). 'Toward a theory of organizational creativity'. *Academy of Management Review*, 18(2), 293–321.

Part II

Employee-Driven Innovation in the Workplace Mediated through Employees' Learning

4
Creating Work: Employee-Driven Innovation through Work Practice Reconstruction

Oriana Milani Price, David Boud and Hermine Scheeres

Considerations of Employee-Driven Innovation generally posit innovation as an advance in the substantive products, services and/or processes of an organization. More broadly, innovation can also refer to anything that seeks to do something new, or address a concern that would not otherwise be met. Employees contribute to innovation in many ways: they can generate and/or implement a product or service; they can generate and/or implement new technologies; however, they can also influence the ways in which an organization adapts and evolves over time in more subtle ways through instigating work practice changes. Although these more subtle changes may not appear under the banner of organizational innovation, they nevertheless contribute to the creation and application of new organizational processes, practices and outputs. They may also never be part of the conscious and explicit agenda of the organization or be something that managers have a strong role in initiating. However, their effects can be cumulative and substantial.

Over the past few decades the relationship between workers and organizations has shifted dramatically. This shift has been underpinned not only by changes in work (e.g. service work, knowledge work) but also in the nature of organizations (e.g. flatter structures, greater autonomy, less formalization, joint ventures to serve global markets). Similarly, contemporary workers not only expect to take responsibility for their own learning, development and careers; they can also expect to take responsibility and ownership for work activities not defined by their organizations or bound to the jobs they were employed to do. We maintain that these changes in the worker–organization relationship and in the nature of work and organizations can open up possibilities for workers to be self-directed and creative – spaces are created where innovation can arise, albeit within certain constraints. We take the view here that innovation arises from the everyday cultural practices of workers – the ways in which workers enact their jobs, interact with each other and seek to become fuller members of their organizations.

It occurs through workers finding ways of meeting their own interests and desires as well as those of their employers.

The focus of this chapter is on how particular forms of employee-driven innovation reconstruct employees' work and organizational practices. It illustrates practices that were not initiated with the goal of innovation in mind, but which have this as a central outcome. It draws on a wider study undertaken in different kinds of Australian organizations of what we termed integrated development practices (IDPs). IDPs are organizational practices whose primary function is the enhancement of organizational effectiveness (Chappell *et al.*, 2009: 359), but which also have a developmental (learning) role. They include practices such as performance appraisal, project work and teamwork. They are practices that facilitate learning, are embedded in work processes, are independent of formal training programs and are not defined explicitly in terms of training and/or education. They are typically managed or implemented by those whose primary job function is not training and learning. Thus, they are not commonly driven by either an explicit learning intention or any explicit intention of innovation.

While many IDPs are familiar and much discussed, others emerged in interviews and observations during our empirical studies. This chapter explores one of the unexpected practices encountered. Recognition of it emerged during fieldwork and interviews, and we termed it 'remaking one's job'. We identified this as a practice in which employees reinvented themselves and their work and enacted this new work in formally reworked jobs in response to their understanding of their employer organization and its changing needs. This chapter examines the dynamics of the processes involved and relates them to the learning of individuals and the organization. These are discussed in terms of the circumstances that afford such possibilities of development and how they can be understood as constituting employee-driven innovation.

We begin by introducing changing ideas about what constitutes innovation and how these changing ideas have brought to the fore the emergence of the concept of employee-driven innovation. Next, we introduce the Schatzkian perspective on practice theory that we have been using to conceptualize our studies and discuss how such notions of practice can frame possibilities for change in organizations. We then introduce a range of employees in two organizations we have studied and exemplify how their practices and what they have taken up from the possibilities of their daily work have created and shaped innovation. Following discussion of how this occurs, we reflect on features of organizations that sustain such innovations.

Innovation and the nature of work

Innovation in relation to work and organizations has been discussed and studied in many different ways. Early understandings of this concept were sustained by an economic focus, where innovation was something new that enabled the creation of value (Schumpeter, 1934). More recent understandings focus on innovation as an organizational process critical for organizational survival: 'the only way a business can hope to prosper, if not survive, is to innovate...to convert change into opportunities...this requires that innovation itself be organised as a systematic activity' (Drucker, 1994: 1) or a process that can be managed. For example, Baumol (2002: 10) describes innovation as 'the recognition of opportunities for profitable change and the pursuit of those opportunities all the way through to their adoption in practice'. Understanding innovation focuses not only on unpacking what is meant by value beyond economic benefit, but also on understanding the processes involved in its production (see, for example, Sundbo, 2003, on strategic innovation; Sundbo and Gallouj, 2000, on loosely coupled systems; Francis and Bessant, 2005, on the 4PModel).

Much research relating to innovation as an organizational process has investigated innovations in the context of research and development (R&D) functions within organizations (Høyrup, 2010). This focus has concentrated on the section of an organization charged with innovation and change as its *raison d'être*. As such, the focus is on a management-driven view of innovation. Research into innovation that has emerged elsewhere in an organization, from employees themselves whatever their primary job function, has not been as widely explored. The concept of employee-driven innovation, as it is termed, is, however, becoming more widely recognized and examined. As has been discussed earlier in this volume, according to Høyrup (2010), employee-driven innovation has a

> focus on innovative practices contributed by any employee (outside the boundaries of his/her primary job responsibilities) at all levels of the organisation...the innovative ideas are embedded in employees' daily work activities – often in working teams – on the basis of their experience and on-the-job training. (p. 149)

Employee-driven innovation begins at the job and worker level and is a bottom-up process that may result in either radical or incremental innovation for products, services, process or markets. Sustaining employee-driven innovation are other organizational processes and managerial activities (Høyrup, 2010). These may include employee suggestion and participation systems, quality improvement approaches or innovative cultural approaches (Teglborg-Lefèvre, 2010). Employee-driven innovation may be conceptualized in terms of a number of elements, including: content (what is new); pace (incremental or radical); recipients (new for whom); value (what is the benefit); process (how it comes about) and drivers (expertise, ideas) (Høyrup, 2010).

It is employee-driven innovation understood as being embedded within daily work activities of job enactment that we focus on here. Following Fenwick's (2003) idea of innovation as a form of 'everyday practice learning', we argue that employee-driven innovation embedded within daily work activities of workers' jobs emerges through their enactments and re-enactments of their jobs. Enactment comprises more than activity or physical 'doings'. To explicate what is meant here by enactment and demonstrate how it occurs and generates innovations, we employ a particular perspective on practice theory. We use this approach to illuminate the ways in which workers enact their jobs and how, in doing so, they can drive innovation within an organization. Practice, we maintain, represents a meso-level of analysis that connects individuals and organization in a mutually constitutive relationship. The notion of 'practice' links the substance of work with the processes involved and with the person engaged in the work. It encompasses the person, the activity and the purpose.

Conventionally, jobs have been understood as reflecting pre-defined roles and tasks nested in an organization's hierarchical configuration and described in documents such as organizational (divisional and workgroup) structure charts, job descriptions, workflows, policies and procedures (Clegg, 1990; Morgeson et al., 2010; Oldham and Hackman, 2010; Rhodes and Price, 2011). This tradition continues in organizations today: workers are recruited, managed and assessed in terms of formally defined roles and responsibilities outlined in organizational documents such as job descriptions, procedures and performance plans (Robbins et al., 2010).

Operating parallel to, and sometimes almost contradicting, these formalized approaches to workers and their work are more informal, everyday ways in which workers think about, construct and reconstruct their work. The survival of contemporary organizations in a rapidly and constantly changing environment demands workers able to envisage and incorporate ongoing change. In this sense all workers are now knowledge workers (Drucker, 1994). They need to understand the whole organization and how they can move beyond the work that has been described and itemized in texts and by managers. Consequently, many of today's workers have greater *de facto* input into the content of their jobs as well as 'considerable latitude to customize...their jobs [make]...changes in the structure and content of jobs...[Employees] do not necessarily have to wait for managers to take the initiative' (Oldham and Hackman: 470). Workers are expected to take greater responsibility and ownership of their work (Child and McGrath, 2001; Grant et al., 2010; Morgeson et al., 2010; Oldham and Hackman, 2010) if they are to survive and flourish in these environments, and if the organization is to be effective. This can be understood as a kind of creative involvement in the organization and thus can position workers as drivers of innovation.

A practice approach

We have used Schatzki's (1996; 2002) and Schatzki's *et al.* (2001) theorizations of practice, as they have proved fruitful in our previous research (e.g. Price *et al.*, 2009) in understanding what workers do at work. This has enabled us to avoid unhelpful polarizations between individual and organization, and between learning and work. Schatzki's framework of what constitutes practices brings together notions of worker, context, relationships and tasks as a means of understanding how workers enact their jobs and at the same time reshape those jobs and the practices of their organizations. In his understanding of organizations as 'bundles of practices and material arrangements' (p. 1863) – as a nexus of existing and altered practices which entwine people, technology and spaces where practices occur – he recognizes that practices frame past, present and future possibilities. Workers enact organizational practices through participation in their jobs and the organization itself. An integral part of the enactment of organizational practices and workers' jobs involves workers drawing on their 'practical intelligibility' (Schatzki 2002: 74). That is, they draw on their understandings of the practices of the organization and they carry out at least some of the activities (doings and sayings) associated with those practices.

While workers' understanding and enactment of practices within a particular organizational context ensure the perpetuation of the organization and its practices, this acceptance does not mean that workers are determined by these practices. Whether consciously or unconsciously, as they enact organizational practices workers draw on their practical intelligibility to carry practices forward and at the same time vary them, enmeshing elements of existing practices with previous understandings of similar practices from other contexts (e.g. previous jobs, prior experiences or knowledge) (Price *et al.*, 2009; Schatzki, 2006). It is through this process of enmeshment of practices that we understand how employee-driven innovations emerge. Workers, as participants in organizations, are active co-constructors and innovators of work and jobs.

Employee-Driven Innovation through remaking one's job

In highlighting the ways in which workers enact their jobs and at the same time innovate within contemporary organizations, we draw on a recent research study. In particular, we utilize data from two Australian workplaces in the public sector: a local government (henceforth *Council*) and a public utility (henceforth *Utility*). *Council* is a local government council in a large metropolitan area. Australian councils represent the third layer of government and are responsible for the provision of services (e.g. libraries, road maintenance, waste collection, building development assessments) and governance at a local community level. *Council* has a hierarchical structure, which includes the elected Council, Mayor, a General Manager, Group Managers, Managers, Team Leaders and 600 employees. *Utility* is an incorporated government public utility organization that provides electricity distribution to parts of the state of New South Wales. *Utility* has a hierarchical structure that includes a State Minister, Board of Directors, Chief Executive Officer, General Managers, Regional Managers and two thousand employees. Our study was conducted in a region of *Utility* that employs three hundred employees.

Council and *Utility* are organizations that have experienced periods of significant change as a result of policy changes at the State (NSW State Government New Public Management Policy) and Commonwealth (National Competition Policy) government levels. For *Council* these changes resulted in new modes of operation. These included greater accountability at the local community level, greater efficiencies in service delivery and resource management, and the adoption of the principles of competition and market contestability. For *Utility*, the application of New Public Management policies, coupled with deregulation and the establishment of a national utility market, significantly shifted the dynamics of this industry. For *Utility* this meant a significant shift in its structures and modes of operation.

The research methods of our project included semi-structured interviews, observations and analysis of organizational documents. Thirty interviews were conducted with workers across hierarchical levels and functions of each organization. We carried out approximately ten hours of observation of work practices. The organizational documents we analysed included annual reports, business plans, policies and procedures, and job descriptions from both organizations. The interviews focused on the retold experiences of workers as they enacted and extended a number of organizational practices pertaining to their jobs. The document analysis provided formalized descriptions of these organizations and enabled us to understand the organizational practices and the jobs within them. The data generated from these methods enabled various accounts of practice and jobs to emerge. Observations of the worksites and work practices allowed us to further understand the work of the organizations and of the work and worker practices. They gave us a 'feel' for each workplace.

Workers in the study enacted the practices of their organizations through their jobs while at the same time changing the practices of their own jobs and their organizations' practices. The workers at *Council* and *Utility* talked about how, in the enactment of their jobs, they varied those jobs and the practices inherent in them – we named this practice *remaking of one's job*. Workers in both organizations remade their jobs by drawing on knowledge, experience and understanding – their practical intelligibility and readings of their changing organizational circumstances. An important feature was that workers sometimes also remade the practices of their organizations. It is this remaking of jobs and organizational practices that may be understood as employee-driven innovation.

To illustrate this remaking, we focus on five employees. *Guy* and *Sally* had been with *Council* for over 15 years and *Stan* had been with them for less than a year. *Miles* and *Harry* were two workers who had been employed by *Utility* for over ten years.

Guy told us that in undertaking his job as Manager, Library and Community Services he drew on his previous work experience in local government in general, and his specific experiences as a Civil Engineer at *Council*. *Guy* told of how he remade his job by undertaking an entrepreneurial approach to community services:

> I manage what is called community development and services – there's a planning policy type area of the unit and there's also service delivery which is the library and also meals on wheels service, immunization service [before I came to this job]...I didn't know a lot [about libraries or community services] but I was active participating in projects that involved [libraries and community services] areas...it was as a given in the engineering areas [where I worked] that you only had to put up a technically difficult report to get the most money – because no one could understand it...I could see the struggle that was happening in the [libraries and community services] areas and I thought how do I move the pot hole to the library or to community development? So I took up that challenge...how do I move that pot hole to the library – I used a similar analogy to try and achieve better resources. (Library and Community Services Manager)

In describing his early experiences as the Manager, Library and Community Services, *Guy* went on to say:

> ...I knew it was going to be difficult...it's more difficult to equate the value of a project of wellbeing of the community or the value that a library may bring to a family over 50 years...[as opposed to a road] and that's always been a toughie to win in terms of traditional senior management structure that if they're economic rationalists...the social entrepreneur [is] really about my view on how I can build a stronger community development area through some entrepreneurial approaches – like moving the pot hole to the library – I've worked with some of the private donation organizations to get some programs up in [this area].

For *Guy*, 'like moving the pot hole to the library' has meant remaking the traditional job of a Manager, Library and Community Services to have a greater focus on talking up the library as a community asset, of bringing to the fore new ways of demonstrating to the 'economic rationalists' its value to the organization. Similarly, in utilizing 'entrepreneurial approaches' and networking with philanthropic organizations from Sydney's wealthier suburbs, *Guy* told us of how he was able to access private donations to fund, for example, a senior women's community group: '...well 10 grand [$10,000] just came along straight up'.

In remaking his Manager, Library and Community Services job to have an entrepreneurial approach, *Guy* was innovating. He was able to create a new way in which the organization could understand the function and role of the library and community services division. Further, in identifying new sources of sponsorship and support for the provision of services to community groups, he was also creating new economic value for *Council* and its community.

Sally, who had worked with *Council* for about 18 years, talked about having been part of *Council* when it initiated its commercial practices. She told us of how she remade her job of Policy Analyst from having a narrow focus on 'developing strategies and policies' to a broader organizational focus of managing the process for developing *Council*'s overall management plan:

> My job title is policy analyst but I have become specialized in creating, developing and producing council's management plan which is [*Council*'s] strategic and operational plan for the future, the strategies which runs about 5 years into the future and the operational runs about 12 months ahead...my main focus is the management plan [process] and what goes into developing it and that includes some heavy consultation, some management workshops, some councillor workshops...the sorts of things you might do in strategic planning and sometimes in change management as well...sort of a aligning the vision and the council's strategies and actions with what's getting done day to day...

In remaking her job to manage *Council*'s management planning process, *Sally* moved beyond *Council*'s previous practices of management planning, which encompassed

> just putting a document together [and telling the operational areas] this is our plan based on the budget, you get the budget and you can see the actions in it and that's what we're going to do and there you have your plan.

Sally focused on developing the management planning practices to encompass internal and external consultation. *Sally* told how her remade job is about

> building relationships with the departments and the managers who have to implement the action plans that the management plan represents and

that means that there's a lot of formal and informal relationship building going and the informal part...the sort of work that you need to do to that isn't always part of your job description...it is the team building stuff you might have and the sorts of meetings that you might have that you do work ancillary...that gives a little bit of relationship building...certainly facilitation...negotiate...manage conflict when there's different priorities and managers are really stressing out about their things and they want them to get done and in the budget.

In remaking her job and the organizational practices of management planning, *Sally* was innovating. In terms of the broader context of innovation, *Sally* had not only reconstructed organizational practices relating to the preparation of the strategic planning document; she also reconstructed the practices of managers involved in this organizational process.

At *Council*, existing workers were not the only workers who were innovating. *Stan*, a newly employed Parking Patrol Ranger, also talked about this practice. In outlining the elements of his job, *Stan* told of how he was responsible for issuing parking infringement notices when drivers were parking contrary to *Council* parking regulations. However, *Stan* approached his job differently; he understood his job as

> not so much giving people tickets, I think for me it's enforcement by presence, I think it made a big difference. I found you can get a lot more people to start doing the proper thing, not so much [by] coerc[ing] them but [by saying to them] 'listen, you can't be here'. They'll go, 'yeah' [because] they don't want a ticket no more than I would want to give people tickets...the people that I've told to move on, they remembered it so the next time'.

Not unlike *Guy* and *Sally*, *Stan* has also been innovating within the general scope of his position. In taking a community education focus to parking patrol, *Stan* has been taking a creative approach in the enforcement of *Council's* parking patrol regulations – he is changing organizational processes and practices pertaining to parking and attempting to shift community cultural values towards parking through education rather than enforcement.

At *Utility* workers were also innovating. *Miles*, a worker who had recently been appointed to a newly created role of Project Manager, explained how he negotiated organizational expectations for project management and existing organizational structural barriers that prevented him from fully implementing project management practices. *Miles* told of how in his new role he was responsible for both the financial and technical management of projects; however, his authority did not extend over the workers who actually carried out the construction of the projects he was responsible for.

In order to address this disconnection between authority and responsibility over projects, *Miles* remade his job to have a greater focus on relationships:

> I don't have anyone that directly reports to me... [the construction crews report to the Operational Managers, who are in a different group] ... I can't demand anything, I can only ever ask... I find that I've got to try and foster relationships... [the construction crews and their managers, the Operations Managers] are key players in the measurement of my output, so I've got to make sure that there is a good relationship between us... that I deal with in those areas well, and understand what their concerns are, if they come to me with a query about some jobs or I need something shoved into the [Construction] programme, having that good relationship helps.

In the remaking of his job to have more of a relationship focus, *Miles* has been innovating the ways in which project management practices could be executed in the context of *Utility* – in a context where project responsibility and authority over project resources remained separate. Through his innovative approach, *Miles* attempted to shift elements of an 'us and them' mentality between the project managers and the construction crews. He reconstructed new organizational practices in order to achieve his own and the organization's asset management outcomes.

Harry, a worker in the same department as *Miles*, described his job as 'Everything Man', a job that

> changes everyday... it's the sort of position you probably put two different people in and you get two very different outcomes... [when I first started] it was just purely feeling your way through... just listened to what everyone was whinging and complaining about [and then I decided to] try and fix one [problem] first... and as you fix one problem, everyone goes oh you fixed that, oh, have a go at this, have a go at that... it's got to the stage where a lot of people now come to me now when they want the process improved, or they want something fixed...

Harry went on to tell us that part of his job was also to fill in when other members of his team went on leave:

> I am able to slot in [and do other jobs] when people take leave, I find while I'm doing their job, instead of just watching someone doing it, [I think there's got to be] a better way... [I put] it in the back of my mind and it goes on my little list [of things to fix] and whenever I get bored, I'll pull that one out and work on [it]... [for example when] the customer services manager left, I virtually slotted into [the] role, which I'd never done before, it was just like, argh... the recording of all the jobs was

very old fashioned, manual filing, no linkages, duplication ... all manual input, and then, the end of the month, month reporting, it was reading through it and then typing it into another one, and paper files ... I went off and rewrote the entire spreadsheet, macros, programming behind it. So basically they're all linked now, so come the end of the month, it doesn't matter where it is, it's one click and you've got your instant figures. Same as you've got an instant report of where everything's up to, every job that's happened.

In his approach to his mostly undefined job, *Harry* has been creating and fulfilling opportunities for innovation. By being able to step into different roles on a temporary basis, *Harry* has been able to develop a comprehensive understanding of the different jobs within his department and the interrelationships among them. In seeking out and fixing problems, *Harry* was reconstructing processes and practices and generating new and innovative ways to execute work within his department.

What can be seen from the accounts of everyday work provided by *Guy, Sally, Stan, Miles* and *Harry* is that these workers have been drawing on their practical intelligibility, their expertise, knowledge and skills to drive new approaches within their organizations. These workers were, in Schatzki's terms, enacting practices that at the same time perpetuated and changed existing organizational practices. They added value to their organizations in various ways, by reconstructing practices and processes, shifting elements of work culture or creating new understandings about the ways in which work could be executed – these workers were innovating in and through their day-to-day work.

While these innovations could from other perspectives be read as examples of individual creativity – they clearly show initiative and personal agency – they arise through more complex convergences and interactions. Innovations need individuals to realize affordances of their work, but they also require the practices of work to provide such affordances. Different kinds of organization and different development practices within them may not provide the possibilities for emergence of innovation.

Sustaining worker innovation

Discourses of change and innovation are heard in the corridors and meeting rooms of all contemporary organizations. Rapid shifts in modes of communication and technologies of all kinds have led to reconceptualizations of what it is to be an organization, what constitutes work, and what it means to be a worker. Organizations can no longer be understood as entities that devise and implement ways of working that persist with adjustments now and again, carried out by workers who come with readymade skills and expertise. Paradoxically, organizational stability incorporates ongoing change, and all employees are expected to be involved in the business of organizing. This involvement includes a shift to greater worker responsibility and trust, rather than control by managers (Child and McGrath, 2001; Drucker, 1994). A structural view of organizations as having fixed arrangements between workers of different kinds and levels is being displaced by one in which, while the imprint of the structure remains, it is continually troubled by a reworking of boundaries and relationships, formally and informally.

Within the organizations discussed here, work occurred in contexts where change had been a key feature. These organizations had gone through periods when goals, values and modes of operating had been significantly reframed. Inherent in these new modes of operation were tensions and gaps between new and old ways of doing things – old ways did not quite fit and new ways were not sufficiently developed to achieve the new organizational imperatives. The context of these organizations was such that workers were afforded the freedom (and perhaps expected) to become the architects in building bridges to fill these gaps, in finding ways to smooth the tensions – these workers were afforded the opportunity to innovate through their everyday work.

As organizations are looking to their employees as trusted organizational members who contribute new ideas and processes, the issue of how to draw employees to this worker and organizational identity, and how to sustain both the generation of innovations and the innovations themselves, is paramount. In the organizations we investigated, we found that employees were able to 'remake' their jobs in an atmosphere in which it was legitimate to do so. They went about creating their jobs, bringing together existing and new organizational practices. In line with Schatzki (2006), these employees were carrying organizational practices forward and at the same time varying those practices in ways that accommodated the situation in which they found themselves. They were able to do so because their organizations valued the outcomes of such innovation and permitted, even encouraged, employees to act accordingly. This is in contrast to older forms of organizations in which fixed or hierarchical relationships are privileged and employees are valued for doing well within their existing job description.

Learning is a necessary and important feature of this kind of innovation, but it is not useful to conceptualize it in ways familiar from education and training. It utilizes everyday learning at work and can be seen as a form of learning in-between, that is, in a third space that is neither personal/social space nor work space, but a hybrid that draws on discourses of both to be productive. This is learning that occurs between 'on-the-job and off-the-job', standing between the formal areas of practice and informal areas of interaction (Solomon et al., 2006: 11). Hulme et al. (2009: 541) suggest that this hybridity pertains to the integration of competing knowledge and discourses; to the reading and writing of subject matter and to the individual and social spaces, contexts and relationships.

Learning is implicated in remaking practices, but there is no direct correspondence between particular kinds and ways of learning and innovation as such. Employee-led innovation occurs through a conjunction of circumstances and the ability of employees to respond to and utilize the opportunities that occur in work. It is not susceptible to being planned by some agent other than those involved. It may best be seen as an emergent practice in which 'learning develops as a collective generative endeavour from changing patterns of interactional understandings with others' (Johnsson and Boud, 2010). That is,

> how workers take up ... opportunities result[s] from a complex combination of situational factors that generate invitational patterns signalled from and by various understandings and interactions among actors doing collective work. (p. 359)

We suggest that it was the openness of job descriptions and the ongoing context of change that enabled these workers to enact their jobs in different ways – they knew change was expected, but what it was and how it could be implemented were open to creative ideas. These expectations and allowances, together with the ways in which *Guy*, *Sally*, *Stan*, *Miles* and *Harry* brought understandings of practice perpetuation and change to their jobs, provided a fertile context for generating innovation. We conclude that it is a coming together of individuals and organizational practices – involving the aspirations of individuals to be part of the organization in ways that are linked more strongly into their desires and interests than the jobs to which they were originally appointed – that is implicated in employee-driven innovations. In these kinds of organizational contexts and approaches to work there is a sense of ownership engendered, rather than a management-devised directive, that we see here helping to drive and sustain innovation.

This chapter has focused on one of the Integrated Development Practices that we identified as part of our wider study. We have chosen to report on this example as it was particularly generative of employee-led innovation. However, it is possible to identify other IDPs in the selected organizations as also generating employee-led innovation, albeit less immediately obvious in their impact. This leads us to suggest that a focus on the practices of organizations and their members is likely to be a fruitful source of innovation data. Such micro-analysis of organizational life illuminates many features that can be obscured in more normative or larger-scale analyses.

Conclusions

Unlike Fenwick (2003), who identified innovation in new enterprises, we have seen substantial employee-led innovation at the everyday practice level in mature organizations. We have identified an interesting set of innovation practices, which is necessarily employee-led. It is one that does not depend on the particular characteristics of an entrepreneurial individual or group, or on altruistic motives in improving the organization. It works through the desire of people to work in ways that satisfy themselves and to effectively meet the needs of whatever kind of work they are involved in. They necessarily utilize existing organizational practices and affordances to do this in similar ways as they use them for other aspects of their work. These development practices may have been created by organizations for quite different purposes – for instance, job descriptions to select applicants and frame employees' work – but, once they exist and are seen as legitimate parts of an organizational infrastructure, they can be used for other ends, including those not foreseen by their instigators.

We have no normative intent in drawing attention to this phenomenon of emergent innovation. It is not necessarily more appropriate, nor may it result in particular kinds of innovation, neither does it necessarily lead to superior kinds of innovation: we have no data on these features and make no claims about these matters. The innovation processes we have identified are ones that cannot be established in advance, or designed into a formal innovation strategy. They are contingent on the engagement of practitioners with the particularities of their own practices and the practices they encounter in their organization. Such engagement is an intrinsic feature of learning work, and with such engagement comes innovation.

References

Baumol, W.J. (2002). *The Free-Market Innovation Machine: Analyzing The Growth Miracle Of Capitalism*. Woodstock: Princeton University Press.
Chappell, C., Scheeres, H., Boud, D. and Rooney, D. (2009). 'Working out work: integrated development practices in organizations'. In J. Field (Ed.) *Researching Transitions in Lifelong Learning* (pp. 175–188). London: Routledge.
Child, J. and McGrath, R.G. (2001). 'Organizations unfettered: organizational form in an information intensive economy'. *Academy of Management Journal*, 44(6), 1135–1148.
Clegg, S.R. (1990). *Modern Organizations*. London: Sage Publications.
Drucker, P.F. (1994). *Post-capitalist Society*. New York: HarperCollins.
Fenwick, T.J. (2003). 'Innovation: examining workplace learning in new enterprises'. *Journal of Workplace Learning*, 15(3), 123–132.
Francis, D. and Bessant, J. (2005). 'Targeting innovation and implications for capability development'. *Technovation*, 25(3), 171–183.
Grant, A.M., Fried, Y., Parker, S.K. and Frese, M. (2010). 'Putting job design in context: introduction to the special issue'. *Journal of Organizational Behavior*, 31(2–3), 145–157.
Høyrup, S. (2010). 'Employee-driven innovation and workplace learning: basic concepts, approaches and themes'. *Transfer: European Review of Labour and Research*, 16(2), 143–154.

Hulme, R., Cracknell, D. and Owens, A. (2009). 'Learning in third spaces: developing trans-professional understanding through practitioner enquiry'. *Educational Action Research*, 17(4), 537–550.

Johnsson, M. and Boud, D. (2010). 'Towards the emergent view of learning'. *International Journal of Lifelong Education*, 29(3), 355–368.

Morgeson, F.P., Dierdorff, E.C. and Hmurovic, J.L. (2010). 'Work design in situ: understanding the role of occupational and organizational context'. *Journal of Organizational Behavior*, 31(2–3), 351–360.

Oldham, G.R. and Hackman, J.R. (2010). 'Not what it was and not what it will be: the future of job design research'. *Journal of Organizational Behavior*, 31(2–3), 463–479.

Price, O.M., Scheeres, H. and Boud, D. (2009). 'Remaking jobs: enacting and learning work practices'. *Vocations and Learning*, 2(3), 217–234.

Rhodes, C. and Price, O.M. (2011). 'The post-bureaucratic parasite: contrasting narratives of organizational change in local government'. *Management Learning*, 42(3), 241–260.

Robbins, S.P., Judge, T.A., Millet, B. and Jones, M. (2010). *OB: The Essentials*. Australia, Sydney: Pearson.

Schatzki, T.R. (1996). *Social Practices: A Wittgensteinian Approach to Human Activity and the Social*. Cambridge: Cambridge University Press.

Schatzki, T.R. (2002). *The Site of the Social: A Philosophical Account of the Constitution of Social Life and Change*. University Park, Pennsylvania State University Press.

Schatzki, T.R. (2006). 'On organizations as they happen'. *Organization Studies*, 27(12), 1863–1874.

Schatzki, T.R., Knorr Cetina, K. and von Savigny, E. (2001). *the Practice Turn in Contemporary Theory*. London: Routledge.

Schumpeter, J. (1934). *The Theory of Economic Development*. Harvard, Massachusetts: Oxford University Press.

Solomon, N., Boud, D. and Rooney, D. (2006). 'The in-between: exposing everyday learning at work'. *International Journal of Lifelong Education*, 25(1), 3–13.

Sundbo, J. (2003). 'Innovation and strategic reflexivity: an evolutionary approach applied to services'. In L. Shavinina (Ed.) *The International Handbook on Innovation* (pp. 97–114). Oxford: Elsevier.

Sundbo, J. and Gallouj, F. (2000). 'Innovation as a loosely coupled system in services'. *International Journal of Services Technology and Management*, 1(1), 15–36.

Teglborg-Lefèvre, A. (2010). 'Modes of approach to employee-driven innovation in France: an empirical study'. *Transfer: European Review of Labour and Research*, 16(2), 211–226.

5
Explaining Innovation at Work: A Socio-Personal Account

Stephen Billett

This chapter proposes a socio-personal account that explains innovations at work. These innovations are central to sustaining the viability of enterprises in the face of continual changes in work requirements and client needs. They also comprise the process through which workers come to both learn and actively remake their occupational practices. That is, innovations have important personal and socially derived purposes, and employee-driven innovations are no exception. Moreover, processes securing and sustaining innovations at work draw on interdependently personal and social contributions. For there to be new practices that secure effective responses to emerging or desired workplace goals, there must be situational premises for these innovations, including the means for them to be supported and adopted, and also the personal engagement by workers for these innovations to be enacted. These personal and social dualistic contributions are used here to provide an explanation of what constitutes innovations at work, in particular employee-led innovations, and how they might progress. Central to this explanatory account are concepts associated with workplace affordances (i.e. how individuals are permitted to participate in work) and individual engagements (i.e. how individuals elect to participate in that work), and, importantly, the relations amongst them. Thus, these contributions comprise a duality that is interdependent and relational. This interdependence includes considerations of the sometimes necessary and sometimes hindering bounding of workers' personal agency by workplace settings.

A socio-personal account of innovation

This chapter seeks to explain what constitutes employee-led innovations through elaborating how the process of innovation occurs within workplaces, and for what purposes. This explanation is advanced through a socio-personal account proposing innovation as comprising two kinds of changes that occur interdependently of each other: i) individuals' learning and ii) the transformation of workplace practices. So, these changes occur in an interdependent relation to each other. Innovations at work also have important purposes in sustaining the viability of enterprises in the face of continual change in work requirements and client needs. So, innovations have salient personally and socially derived purposes, and employee-driven innovations are no exception. Moreover, the processes that secure and sustain innovations at work also draw interdependently on personal and social contributions. For there to be new practices that secure effective responses to emerging or desired workplace goals, there must be both situational premises for these innovations and the means for them to be supported, and also personal practices of workers so that these practices can be engaged with and enacted. The personally and socially dualistic account of innovations presented here offers an explanation of what constitutes innovations at work and how they might progress, and, in particular, how employee-led innovations might be understood and promoted. Innovations at work, then, because they arise at the personal and social level, comprise the range of personal and/or situated novel actions that can arise through work, including developing new products and offering new services, and improvements to and efficiencies in work practices. Moreover, because we are concerned here with employee-led innovations, these activities extend to workers generating new practices and ways of progression and other decisions associated with exercising their personal and/or collective influence in the workplace, such as protecting or extending their interests and influence (e.g. gaining control of work activities) and even subversion.

Therefore, to understand the nature, forms and enactment of these changes, both social and personal premises and factors need to be considered. Hence, the explanatory concept advanced here is referred to as a socio-personal account. In addition to this is the relational interdependence (Billett, 2006) between these two sets of premises and factors that stand as bases to explain what constitutes innovations at work and, in particular, how employee-driven innovations can be conceptualized and appraised. This account proposes the key premises for the personal learning and cultural practice of remaking work and its transformation (i.e. innovations) as comprising an interdependent and relational process and the outcome of employees' engagement in work-related activities. It is through these activities that employees learn and new practices are enacted. Importantly, the account elaborates how everyday processes of thinking and acting at work are constructive acts through which work tasks and process are reconfigured in response to new requirements and to specific situational requests or problems (e.g. work tasks). This everyday process of work-related thinking and acting both ordinarily and necessarily leads to the remaking of occupational practice and, by definition, comprises employee-led innovations, as well as new learning for those employees (Billett et al., 2005). Hence, when employees confront and respond to significant changes to practice (i.e. innovations), new learning for those enacting these tasks and the transformation of the practice co-occur. It is likely that most of these actions require and bring about change in individuals (Rogoff and Lave, 1984). As individuals engage in activities, a legacy arises in terms of either the reinforcement and refinement of what individuals already know or the extension of what they know (Lave, 1993). A key point here is that innovations at work constantly occur as individuals engage in their everyday work-related activities and respond to emerging challenges and new requirements for effective work performance. Importantly, through engaging in their everyday activities these workers are also participating in remaking their occupational practice (i.e. remaking culture).

It follows, then, that, regardless of whether innovations are intentionally organized (e.g. as with new products or work processes) or arising through everyday work activities (e.g. changing client base, new materials, emerging technologies), the transformation of occupational practices arises in ways shaped by how employees engage in and learn through activities that are to various degrees new to them. Thus, innovations are as much about those individuals' learning as they are about the implementation of new practices, and one is unlikely to occur without the other. Moreover, these new practices are likely to become increasingly more frequent and potentially greater in scope as products and production cycles shorten, leading to continual changes in products and processes (Bailey, 1993b). Indeed, it is often claimed that companies that intentionally aim to continuously improve their processes and products, and are directed towards these purposes, are those most likely to be able to respond to the kinds of emerging challenges which are not always predictable or foreseeable, thereby securing their continuity as viable workplaces. Hence, there is interdependence between employees' learning, on the one hand, and the remaking of practice of the workplace, on the other, that works in the interest of both workplace and workers. Consequently, as long argued in the human resource development literature (e.g. Carnevale, 1995; Howard, 1995), intentional efforts to secure innovations need to include a consideration of employee engagement and development, and a workforce's capacity to respond to predicted and unanticipated changes, all of which are likely to be founded on workers' capacities to adapt to and enact innovations. Hence, employee-led innovations are likely premised on the skilfulness and adaptability of the workers who actually engage in and realize the transformation of practice, and also the degree to which those workers are invited to exercise their discretion. Constraints on that discretion (i.e. bounded agency) might be legitimate, to care for and maintain long-proven practices, or, alternatively, to protect existing interests. Examples from a body of research are drawn upon here to elaborate and advance these understandings.

In all, it is proposed that both ordinary and innovative practices have to be employee-driven, because it is workers who enact work tasks, confront new challenges and respond to those new tasks. In both cases, the enactment of employee-driven innovations can be understood through a process of bounded agency (e.g. Shanahan and Hood, 2000), that is, the degree to which employees are able to exercise their agency, within the constraints provided by the workplace. The exercise of unbounded individual agency can lead to inappropriate, unhelpful and potentially disastrous outcomes for the workplace, and possibly workers. Hence, there are likely to be constraints, shaped by workplaces, within which the processes of engaging in and remaking of work occur. These constraints extend to what constitutes innovations at work and when such innovations are welcomed, or not, in workplace settings. However, the exercise of agency can also lead to changes that transform work practices.

This case is advanced through first discussing what constitutes innovations from perspectives that account for both personal and social factors, and the relationships between them, providing a framework to explain what constitutes employee-led innovation. The discussion then turns to consider issues associated with employee-driven innovations, and how these progress in workplaces. It is here that the concept of bounded agency at work is discussed to account for the prospects for such innovations.

Constituting innovation

The majority of activities in which humans engage are goal-directed. That is, they are intentionally directed towards achieving particular goals, for example to quench thirst, overcome hunger, or move from one place to another, and, of course, there are goals relating to their engagement in paid employment. It is likely that the workplace goals to be achieved are most often premised on the needs of others (e.g. employers, owners, managers, clients, customers), as are the means by which these intentions are supposed to be realized and judgements about how well these goals are being met. Yet, they can also be about individuals positioning themselves, finding ways to advance their personal or collective interests or secure their own work life trajectory (Billett and Pavlova, 2005). So, workers, even when self-employed, are often responding to intentions established by others in the conduct of their work. Of course, the influence and control of these others – the suggestions of social partners – will differ widely depending upon the kind of work being undertaken (i.e. occupation), how that work is being practised (i.e. workplace organization), and the role and status of those undertaking the work (i.e. their standing in the workplace). Clearly some forms of work (e.g. air traffic control) are more regulated, controlled and monitored than others. However, and in some ways regardless of this, the conduct of work is also premised on how individual workers construe, construct and enact their work activities, for which others (e.g. managers, supervisors etc.) may well have set goals and will also monitor these workers' performance, and make judgements about and constrain their activities. These work activities include establishing a specific goal for the task. For instance, whether the worker is a nurse, doctor, engineer, hairdresser, vacuum-cleaner salesperson or lighthouse keeper, they are likely to establish the goals required to meet the needs of the work task. Even the most tightly supervised workers will probably be able to exercise some form of discretion in their work practices (Billett *et al.*, 2005). No amount of prescription can wholly control the work of individuals, such as those who teach students, because they are engaged in activities that require moment-by-moment decision-making in response to particular problems in particular situations and at particular points in time. To illustrate this point, and referring to teachers' work, Schwab (1983: 245) states that

> Teachers practice an art. Moments of choice of what to do, how to do it, with whom and at what pace, arise hundreds of times a school day, and arise differently everyday and with every group of students. No command or instruction can be so formulated as to control that kind of artistic judgement and behaviour... teachers must be involved in debate, deliberation and decisions about what and how to teach.

Indeed, possibly no amount of control or surveillance can ultimately shape how we come to value or desire something (Foucault, 1986) and, hence, construe and respond to what we encounter. Wertsch (1998) makes the distinction between mastery and appropriation. The former is the process of superficial compliance to and apparent acceptance of what others demand or want, whereas the latter is when individuals willingly assent to accepting what others suggest, because it is consistent with their beliefs and values. Yet, regardless of this, even when seeking to faithfully follow existing procedures and secure outcomes, as in conscious attempts to appropriate their worth, it will be necessary for individuals to exercise discretion, simply because these expectations and requirements to follow others may not be easily, clearly or unambiguously understood. Here, Newman *et al.* (1989) helpfully remind us that if the suggestions of the social world were clear and unambiguous there would be no need to communicate (i.e. talk), because suggestions, such as requests from others with whom we work, would require no further clarification or elaboration. But these suggestions are not always clear and unambiguous, and as a consequence require individuals to actively construe and construct meaning from what has been suggested. This construction is premised on their ways of construing it, which Valsiner (1998) refers to as the cognitive experience, shaped by earlier or pre-mediate experiences. Typically, these processes are initially directed to secure intersubjectivity – some shared level of socially shared meaning – before shaping how individuals elect to engage in work activities. Yet, at the centre of this process is the personal process of meaning-making.

So, ultimately, and within the constraints of the work practice, it is individuals who assent to and decide how they undertake a work task, including the degree of effort and finesse that they exercise when undertaking that task, and how and to what degree they will monitor its enactment. This is not to say that others will not and have not influenced the organization and conduct of that work task. Yet, how the individual considers the task (i.e. conceptualizes it) and goes about completing that task (i.e. selects and utilizes procedures) and how they think about and value it are likely to be shaped by their personal dispositions and preferences. This process necessarily engages them in exercising their personal construals, preferences, values and conceptualizations of the task and how it might be best enacted, given the constraints and the goals. For instance, when confronted with the same problem tasks, hairdressers responded in quite distinct ways to undertaking those tasks. Some responses were shaped by the requirements of the hairdressing salons in which they worked. However, the detail of establishing the sub-goals and processes for the task and then the enacting of those tasks arose from their particular preferences and practices (Billett, 2003). That is, these workers ultimately establish the specific goals to work towards, select the means to realize those goals and then work towards achieving them. Therefore, although the social and physical worlds, such as those comprising workplaces, may provide the activity and the context for its enactment, suggest how it should be undertaken and for what purposes (i.e. provide the goals), and even seek to enforce the processes for engaging in activities, the conceptual and procedural means to complete it are premised within individuals' decision-making. This is not least because these work tasks often have particular dispositional requirements (e.g. thoroughness, speed, care, politeness). That is, they are largely premised on human intentionalities. Hence, how individuals construe, construct and enact these tasks is what ultimately shapes how they are completed, and likely in personally idiosyncratic ways (Valsiner, 2000). So, in many ways, everyday work activities *per se* can be thought of as being employee-driven, albeit in situations where there are different kinds of prospects for employees to exercise their agency, including the kinds and extent of discretion they are openly able to exercise. Therefore, beyond considerations of possible options for the exercise of discretion, there are likely to be constraints in terms of what employees can actually implement in their workplaces.

Much of what has just been discussed applies equally to activities that are, on the one hand, either intentionally imposed innovations, such as new products, materials or technologies, or, on the other, new changes at work that are advanced by employees themselves – employee-driven innovations. Certainly, with such innovations, the imperative for change may well come from others, or alternatively from the employees themselves. Yet, regardless of their source, given their personally premised processes, it will be individuals who are ultimately central in realizing how these innovations are advanced, shaped to be most effective and enacted in practice. Beyond the constructive processes associated with enacting work activities that are already known about (i.e. routinized activities), which are still subject to the individual's construal and enactment, workplace innovations (i.e. non-routine or novel activities) make new demands on the employees who have to implement them, not least because these individuals have to learn about them and how they should be enacted. Again, as these learning processes are undertaken by individuals, and likely in personally specific ways, much of those processes are premised upon what individuals already know, making those processes personally specific in some ways.

Consequently, doing something new (i.e. novel) requires individuals to learn about it, and doing this learning necessarily contributes to their ontogenetic development – personal development across their life histories. Because of the uniqueness of individuals' ontogenies, the degree to which the innovation and learning are novel to the individual is likely to be person-dependent (Billett, 2009). What for one individual is an entirely new work task, for others is to varying degrees an adjustment to what they already know, can do and value in some way. The difference here is between what might be described as new learning, or non-routine problem-solving, and mere refinement or routine problem-solving. The former (non-routine) problem-solving and learning likely requires individuals to extend their concepts, procedures and dispositions to accommodate the new task (Anderson, 1993; Groen and Patel, 1988). This cognitive process is effortful and demanding because it requires developing these forms of knowledge further (i.e. creating new knowledge) (Tobias, 1994). Such processes may well require individuals to develop new categories, cognitive structures or capacities (Fitts, 1964). Because these kinds of knowledge may well be quite demanding to learn, their learning will likely be premised on what individuals already know, and the degree to which they engage in the effortful process of learning, refining and honing that knowledge will be central to the degree to which they learn that knowledge. That is, they are all premised on personal factors of prior knowledge, and the exercise of engagement and effort, which are premised in their dispositions.

Therefore, engaging in practices that are novel to individuals is likely to be effortful, and this effort will only be extended if individuals believe them to be worthwhile enough for them to expend the effort required for that learning. Hence, the degree to which these initiatives are seen as being worthwhile by employees is likely to be central to how they are enacted in the workplace. For instance, studies of individuals engaging in innovative practices indicate that, unless they have the interest and develop the capacities to successfully implement such practices, they will only engage with them in a cursory way (i.e. as in mastery), and likely even then only when supervised and monitored. For example, one of the classical studies of teachers' professional development found that, unless teachers had not only developed the technical capacities to implement new practices but also enjoyed success with their implementation, they would not develop a commitment to these practices (Mclaughlin and Marsh, 1978). Hence, when the support provided to teachers was withdrawn and supervision was not maintained, teachers who had not experienced success with the new practices did not continue with the proposed innovations. That is, their appropriation of these ideas was premised on the successful use of those strategies. Only then would they appropriate them or take them to themselves as Luria (1976) suggests. Of course, in other kinds of work where performance can be closely monitored and supervised, it might well be claimed that it would be necessary to enact the innovation regardless of the workers' preferences.

Noteworthy in the study referred to above was the finding that who initiated the innovative practice was less important in terms of employees' adoption of it as a worthwhile practice than whether the teachers became committed through experiencing success when enacting the new practice (i.e. their appropriation of it). However, uninformed, uncommitted and unconvinced workers are unlikely to engage in activities effortfully and conscientiously, and may struggle to enact processes and secure goals they do not understand even when closely supervised. Hence, the processes for them to enact innovations are strongly mediated by personal factors that themselves have been shaped by earlier socially derived experiences, including how they elect to assent to these changes. Yet, even in these situations, workers may well innovate in different ways, and maybe not as intended by those who permitted them to bring about change. That is, variously, they may either improve on the innovation, adapt the innovation in ways which suit their work or work team, or undermine an innovation introduced by management or owners so that it will be withdrawn. Even the most willing of workers, who are keen to implement initiatives with great fidelity, may not be able to enact them unless they have the capacities to do so, or the understanding of how to progress. The point here is that, either collectively or individually, the conduct of innovations will be premised upon individuals' capacities and interest to engage with them, albeit within the constraints of their supervised circumstances. That is, individuals are likely to exercise personal agency associated with their current capacities, understandings, preferences and energy. This application of personal discretion has been referred to as bounded agency, which is described as 'the dynamic interplay between individual effort, group-based strategies and social structures' (Shanahan and Hood, 2000: 131). Here, it is captured as the exercise of personal agency within the constraints of a particular social circumstance.

There are, of course, other situations where workers are encouraged to innovate, and this, indeed, is part of their work role in responding to client or workplace requests. Yet, these workers may well confront other kinds of bounded agency. That is, they will be able to be innovative within the constraints of what is permissible within the workplace setting. For instance, in a study of hairdressers across four hairdressing salons

(Billett, 2001), it was noted that the scope of possible hair treatments (e.g. haircuts, colouring, etc.) was very much shaped by the 'practice of the community' (Gherardi, 2009) in which they participated. That is, there were sets of hairdressing practices that were aligned with what occurred within the particular salons ('what we do here is ... '). Regardless of whether the salon's clientele comprised lonely old ladies, people in a rural community or clientele from an inner-city upscale suburb, there was an overall prescription of the kinds of hairdressing practices that were performed in each of those salons. These practices in some ways constrained the kinds of hairdressing treatments that could be provided. Occasionally, this prescription was overridden, but inevitably by either an owner or a manager, or a self-employed hairdresser. Yet, in these salons, as in other workplaces, there is a need for change and innovation for these workplaces to be sustained, and there is also the need for individuals to remain skilled and competent to sustain their employment. So there is often, but not always, an interdependence between the kinds of changes (i.e. innovations) that workplaces want to bring about and the kind of learning that individuals need to engage in to maintain their employment. Consequently, innovation is something that needs to proceed interdependently, relying upon and being enacted on the basis of imperatives for workers, and to be tolerated in the workplace. When viewed in this way, the process of initiating and realizing innovation in workplaces needs to be premised upon developing the capacities and interests of those who are to adopt and enact the new practices. That is, changes to individuals (i.e. their learning) are likely to be central to bringing about all the adoption of innovations in workplaces.

There is also one further and salient consideration here. That is, when individuals are engaged in new activities and practices at work, they are also engaging in the process of remaking cultural practices (Billett *et al.*, 2005). So, as individuals learn (i.e. engage in innovative practices), they are participating in a process of sustaining, reproducing and transforming those practices. The remaking of those practices can comprise the continuation of practices that have arisen through cultural need, have been reshaped over time and find manifestation in particular circumstances across generations. Thus, individuals are engaging in a process of continuity and transformation of their occupational practices. Yet, given that the requirements for work are constantly changing, and occupational practices with them, the process of remaking likely involves advancing and potentially transforming the occupational practice to meet these needs. The point here is that workplace practices are dynamic and constantly needing to respond to existing and emerging work requirements, and through employees' actions the work practice is constantly being remade. Yet, this remaking is also premised upon workers' interests, capacities and intentions. Consequently, to explain the process of enactment and its outcomes, these activities, being either personal or situational, need to account for both a combination of and relationships amongst both sets of factors. Given the interdependence between the workplace as a social setting and the worker as an active learner, it is necessary now to attempt to explain how innovations at work likely progress and might be supported. This element of the socio-personal account of innovations is the focus of the next section.

Innovations at work

Following on from the above, the development, adoption and enactment of innovations by employees can be conceptualized as progressing through a duality that comprises, on the one hand, the affordances of the workplace and, on the other, individuals' engagement with innovative practices. Workplace affordances comprise the invitational qualities of the workplace: the degree to which individuals are invited to participate and are supported in their participation. Some workplace affordances are very high, offering individuals opportunities for full engagement, and are highly supportive in terms of opportunities for learning, further development and advancement, and the exercise of occupational discretion. Other workplaces may be less invitational, being restrictive and limited in their support for workers. Further, the same workplace may offer different levels of affordances to workers. For instance, in his study of a computer manufacturing company in America, Darrah (1996) found that, whereas the engineers in the workplace were given great discretion, championed as being innovative and afforded high levels of remuneration, almost the opposite was true of the production staff. Yet, it seems that the production staff had to engage in flexible and innovative work practices to ensure the production schedules of the workplace could be met. These practices included monitoring and self-organizing the rearrangement of production schedules based upon the availability of component parts. Put simply, they had to innovate to achieve the goals of their everyday work activities. Consequently, an objective assessment of the kinds of work that were being undertaken suggests that both kinds of workers engaged in innovative work practices, and both were employee-driven. However, affordances in the form of discretionary work practices, recognition, opportunities for development and remuneration were quite asymmetrical across this workforce. Here, while it was quite likely that the electronics engineers in this company would be acknowledged and rewarded (i.e. championed) as engaging in employee-driven innovations, the same would not be said about the production workers. Therefore, when considering what constitutes innovation it may be necessary to look beyond the kinds of occupations that are held to be innovative. In different ways, workplaces will afford opportunities to workers that will support their discretion in being innovative or, alternatively, will seek to manage a process to bring about changes in ways which may be quite restrictive, sometimes with good reason. For instance, in meeting specific requirements, safety standards or tolerances, or where potentially risk-filled outcomes could occur, those affordances might be very carefully monitored. Hence, there is a need to consider the boundaries within which employee agency might best be exercised in generating and enacting innovations. To take an example, air traffic controllers might well be afforded discretion in some parts of their work, yet those aspects associated with the routing of aircraft would be highly constrained under most circumstances, and even then within prescribed flight regulations. It is in these situations that employee-derived innovations are not likely to be welcomed, because of the risks and under-considered threats to existing and safe practices that they might generate. Yet, included in this consideration of innovations are the ways that individuals themselves will respond to positions, acting in particular ways that are premised on meeting individuals' needs more than their workplace.

Of course, some workplaces will deliberately and strategically encourage employees to engage in promoting innovations. These are utilizing employees both to secure improvements in the quality of service and the production of goods, and also to secure a greater engagement with the workplace. This approach is often tagged as the human resource management kind of strategy that was born out of the Hawthorne experiments, which showed that employee engagement can lead to significant increases in productivity. Following from this consideration of engagement, it has been suggested that approaches to organizing work need to account for three key factors: i) motivation, ii) employee skills and iii) organizational structures (Bailey, 1993a). Another kind of evidence suggests that one of the qualities of small-to-medium-size enterprises which have stood the test of time is a high level of employee engagement (Rowden, 1995). Other researchers suggest that this kind of workplace approach should be taken because it is ethically correct (Carnevale, 1995; Howard, 1995). Yet, although much referred to within the literature, it seems that, even where strategies of this kind have been trialled, many organizations either have never bothered to fully implement this kind of strategy or, indeed, have retreated from providing workers high levels of discretion, because the evidence for its effectiveness remains quite limited (Bailey, 1993a). In reflecting upon this trend, Bailey proposes that a full understanding of what factors are likely to engage workers, and direct and sustain their efforts towards improving the workplace, is required. It is proposed here that the interdependence and relations between personal and societal factors comprise such factors.

Given all this, the concept of affordances – the ways in which individuals are invited to participate in the workplace – may be helpful in explaining the ways in which employee-driven innovations might best be engaged. However, related to the ways workplaces might effectively engage with innovations are the issues of how workplace errors are responded to and dealt with, and the degree to which they are acknowledged and can promote innovations in workplace practices. Quite likely, in workplaces where innovation is highly prized, there will be inevitable errors as workers engage in practices that carry certain risks, because they are dealing with ideas and practices that may not have been proven over time. Such workplaces will likely require a degree of tolerance of errors, ambiguities and failures (Bauer and Mulder, 2007) as this is the kind of environment in which innovations can be seen as welcomed and potentially important components of individuals' work. Moreover, more generally, when dealing with non-routine situations that are likely to be generative of new practices or responses to emerging problems, this tolerance is likely to be a helpful quality. Of course, as with air traffic controllers, potentially dangerous activities would need to be highly constrained, particularly where the consequences for others can be dire (e.g. patients in hospitals, children in schools). In short, workplaces that intentionally seek to promote employee-directed innovations would need to afford the kind of opportunities and environment in which novel ideas could be advanced, trialled and evaluated, whilst protecting the proposers from sanctions should these efforts turn out to be unhelpful or unproductive.

However, the other dimension of the duality of affordances and engagement needs also to be considered here. That is the degree to which individuals take up the invitation either to be innovative or to conform to existing workplace practices. All of this can play out in quite distinct ways, including the over exercise of personal agency. For instance, in one study an individual who was by trade a fitter, but was employed as a production worker, exercised his agency in innovative ways which were not invited by the workplace (Billett and Boud, 2001). Being keen to secure work as a fitter (i.e. trades work), he engaged with his work and learning about that work in very agentic (i.e. highly proactive and personally directed) ways. This agency included him following the workers, observing them physically and listening into their intercom conversations, and then asking them how they went about their tasks and asking for justifications for their approaches to their work tasks. This innovative and agentic behaviour led him to questioning trades workers during their lunch breaks, and also following them into areas which were outside his safety limits. Ultimately, he was sanctioned for his agency, as this behaviour was seen to exceed safe and authorized work activities, and was an irritant to other workers. In another study, similarly, a worker also adopted innovative and agentic practices, as a result of being sent on a training course, which, again, were not welcomed by the workplace because they contravened workplace norms and practices (Billett and Pavlova, 2005). Again, this very agentic worker was sanctioned for exercising excessive agency, which included him advising supervisors and senior management that he could assist them to perform their work roles more effectively if they would but ask him for advice. This personal discretionary basis for engagement can also play out in quite the opposite way to the two examples above. Hodges (1998) reported withdrawing from an early childhood education course when the values and activities she was being taught and expected to implement in kindergartens conflicted with her beliefs about how children should be viewed and treated. All of this suggests that, whilst employee-led innovation is occurring all the time, there is a need both to promote the active engagement of employees in the intentional process of innovation in order to achieve the goals of the workplace, and to extend the learning of workers. However, the promotion and exercise of this agency may well need to be curtailed, or at least managed in ways such that it does not have a deleterious effect on either the short-term or long-term viability of the workplace. Hence, next, and finally, it is necessary to turn to the concept of bounded agency and how this concept can be used to understand the organizing and shaping of the prospects and scope of employee innovation. It also provides a way of explaining how the relationships between the requirements of the social practice (i.e. workplace) and the exercise of learners' (i.e. workers') agency come to be negotiated, and also transformed through these negotiations: how the combination of social and personal contributions is mediated.

Innovations and bounded agency

As has been discussed above, the exercise of employee-directed innovations occurs continuously and ordinarily as workers engage in their everyday work activities. This engagement likely generates the remaking and transformation of workplace practices. So, innovative activities occur through the conduct of everyday work activities. As noted, there are also intentional efforts on the part of owners and managers to secure workers' engagement in making particular kinds of innovations to, for example, improve workplace productivity or identify new goods and services and new practices. These kinds of intentional efforts may well acknowledge that those who undertake work tasks may be positioned most effectively to identify opportunities for innovation and enact those innovations, not least because of their interest and familiarity. Usually associated with this kind of initiative is the granting of some degree of discretion to workers so that they can innovate. This discretion might simply be a tolerance for suggestions about change. As has been noted above, the level of discretion in work roles may be increasing through the kinds of work which are currently being enacted in countries with advanced economies – professional, paraprofessional and high-skill work. Even in these types of occupations discretion is likely to be constrained by employment and organizational boundaries. That is, there are limits to discretion and the effectiveness of practice that individuals are able to generate and exercise without the permission of others. Most of these limits can be seen as boundaries around the potential agency of workers – bounded agency (Shanahan and Hood, 2000). This boundedness pertains as much to professional as to other forms of work, given the increased emphasis on regulation, risk management and issues of liability, which shape the conduct of many forms of employment. In different ways, there are likely to be constraints upon workers of all kinds that are shaped by legislation, mandation or practice requirements, and so on (Billett et al., 2005). Indeed, even those who practise in relative privacy, away from the observation of others (e.g. teachers) are increasingly subject to scrutiny and constrained by external mandates of how they should be conducting their work (e.g. what they ought to be teaching and how). In these circumstances, it is noteworthy that most forms of employment arise because there is a need for them, usually generated by others.

Moreover, it is these needs and how they transform over time that shape what constitutes key imperatives for the practice of the occupation. To continue the above example, few teachers ever establish their own schools, based on their particular philosophy; instead, most are employed in educational institutions that are government, religious or privately sponsored. These institutions exist because they address particular societally derived and identified needs. Although teachers can and do resist and ignore these boundaries to some degree, their actions are nevertheless in some ways constrained by them. Elsewhere, these boundaries are the means by which work is organized and undertaken, and these can be central to what constitutes particular occupational practices and the demarcations that surround them. Indeed, often such demarcations are the very premises for occupational practice. Hence, innovations that seek to tamper with such demarcations might be resisted and rejected. Whether changing the practices of medical practitioners or retail workers, such innovations may likely be countered.

So, there are boundaries to the generation and adoption of innovations that need to be considered in employee-driven innovations, and these will exercise individuals' agency in particular ways. First, there is the degree to which such innovations pose a risk to the workplace in terms of its viability and continuity or have the potential to promote those outcomes. This risk will comprise a consideration of how the innovation will be welcomed by the workplace and also the scope of the proposed innovations. Second, the potential impact and consequence for the practice of the workplace community likely needs to be accounted for in both the approach to and the enactment of any innovations. This accounting includes a consideration of the potential scope of the impact of the innovation. Consequently, there are likely to be boundaries that serve to limit both the prospect and the scope of employee-driven innovations, for reasons associated with sustaining the practice and preserving the practitioners as much as bringing about changes to practice and practitioners.

In sum, proposed here is an explanation of how innovation at work progresses through an account that comprises the contributions of, independence of and relations between social and personal contributions to thinking, acting and learning at work. In all, employee-driven innovations comprise negotiations between these factors. These interactions between personal agency and social norms and practices are enacted through these negotiations, and, reciprocally, both that agency and those practices are developed through these processes. That is, individuals' agency is refined, extended or honed through these processes, and the practices that comprise paid employment are also remade and, potentially, transformed. Importantly, innovations at the personal level comprise new learning that likely arises from some level of effortful engagement and the generation of a response that is in some way novel to the individual. These premises can be extended to innovations that are directly aligned to the workplace's goals, or those of the individual learner, worker and innovator.

References

Anderson, J.R. (1993). 'Problem solving and learning'. *American Psychologist*, 48(1), 35–44.
Bailey, T. (1993a). *Discretionary Effort and the Organisation of Work: Employee Participation and Work Reform Since Hawthorne*. New York: Teachers College, Columbia University.
Bailey, T. (1993b). 'Organizational innovation in the apparel industry'. *Industrial Relations*, 32(1), 30–48.
Bauer, J. and Mulder, R.H. (2007). 'Modelling learning from errors in daily work'. *Learning in Health and Social Care*, 6, 121–133.
Billett, S. (2001). 'Knowing in practice: re-conceptualising vocational expertise'. *Learning and Instruction*, 11(6), 431–452.
Billett, S. (2003). 'Sociogeneses, activity and ontogeny'. *Culture and Psychology*, 9(2), 133–169.
Billett, S. (2006). 'Relational interdependence between social and individual agency in work and working life'. *Mind, Culture and Activity*, 13(1), 53–69.
Billett, S. (2009). 'Personal epistemologies, work and learning.' *Educational Research Review*, 4, 210–219.
Billett, S. and Boud, D. (2001). *Participation in and guided engagement at work: workplace pedagogic practices*. Paper presented at the 2nd International Conference on Learning and Work, Calgary, 26–28 July.
Billett, S. and Pavlova, M. (2005). 'Learning through working life: self and individuals' agentic action'. *International Journal of Lifelong Education*, 24(3), 195–211.
Billett, S., Smith, R. and Barker, M. (2005). 'Understanding work, learning and the remaking of cultural practices'. *Studies in Continuing Education*, 27(3), 219–237.
Carnevale, A.P. (1995). 'Enhancing skills in the new economy'. In Howard, A. (Ed.) *The Changing Nature of Work* (pp. 238–251). San Francisco: Jossey-Bass Publishers.
Darrah, C.N. (1996). *Learning And Work: An Exploration in Industrial Ethnography*. New York: Garland Publishing.
Fitts, P.M. (1964). 'Perceptual-motorskill learning'. In Melton A.W. (Ed.) *Categories of Human Learning* (pp. 243–285). New York: Academic Press.
Foucault, M. (1986). *The Care of the Self: the History of Sexuality* (Vol. 3) (R. Hurley, Trans.). Harmondsworth: Penguin.

Gherardi, S. (2009). 'Community of practice or practices of a community?' In Armstrong, S. and Fukami, C. (Eds) *The Sage Handbook of Management Learning, Education, and Development* (pp. 514–530). London: Sage Publications.

Groen, G.J. and Patel, P. (1988). 'The relationship between comprehension and reasoning in medical expertise'. In Chi, M.T.H., Glaser, R. and Farr, R. (Eds) *The Nature of Expertise* (pp. 287–310). New York: Erlbaum.

Hodges, D.C. (1998). 'Participation as dis-identification with/in a community of practice'. *Mind, Culture and Activity*, 5(4), 272–290.

Howard, A. (Ed.) (1995). *The Changing Nature of Work*. San Francisco: Jossey-Bass Publishers.

Lave, J. (1993). The practice of learning. In Chaiklin, S. and Lave, J. (Eds) *Understanding Practice: Perspectives on Activity and Context* (pp. 3–32). Cambridge, UK: Cambridge University Press.

Luria, A.R. (1976). *Cognitive Development: Its Cultural and Social Foundations*. Cambridge, Massachusetts: Harvard University Press.

Mclaughlin, M.W. and Marsh, D.D. (1978). 'Staff development and school change'. *Teachers College Record*, 80(1), 69–94.

Newman, D., Griffin, P. and Cole, M. (1989). *The Construction Zone: Working for Cognitive Change in Schools*. Cambridge, UK: Cambridge University Press.

Rogoff, B. and Lave, J. (Eds) (1984). *Everyday Cognition: Its Development in Social Context*. Cambridge, Massachusetts: Harvard University Press.

Rowden, R. (1995). 'The role of human resources development in successful small to mid-sized manufacturing businesses: a comparative case study'. *Human Resource Development Quarterly*, 6(4), 335–373.

Schwab, J. (1983). 'The practical 4: something for curriculum profesors to do'. *Curriculum Inquiry*, 13(3), 239–265.

Shanahan, M.J. and Hood, K.E. (2000). 'Adolescents in changing social structures: bounded agency in life course perspective'. In L.J. Crockett and R.K. Silbereisen (Eds), *Negotiating Adolescence in Times of Social Change* (pp. 123–136). Cambridge, UK: Cambridge University Press.

Tobias, S. (1994). Interest, prior knowledge, and learning. *Review of Educational Research*, 64(1), 37–54.

Valsiner, J. (1998). *The Guided Mind: A Sociogenetic Approach to Personality*. Cambridge, Massachusetts: Harvard University Press.

Valsiner, J. (2000). *Culture and Human Development*. London: Sage Publications.

Wertsch, J.V. (1998). *Mind As Action*. New York: Oxford University Press.

6
Innovation Competency – An Essential Organizational Asset

Lotte Darsø

Ask any top manager whether he wants innovation in his company and he will certainly say yes. Ask the same question further down the organization and most managers will respond in the same way. Follow these managers closely and most of them will have their attention on anything but innovation. This type of hypocrisy is well known in organizations (Brunson, 2003) and, in fact, it can easily be explained. Managers have many other responsibilities and goals that must also be met, such as optimizing production, increasing sales, ensuring quality, and so on, which means that finding the extra time for innovating can be difficult. So, on the one hand, too much creativity and innovation would be a nightmare for managers (Levitt, 2002); on the other hand, creativity and innovation have become crucial for companies to survive. What to do about it?

In most companies, the solution is to have a special department for innovation activities and new business development. Appointed people, who have core knowledge about the products of the company and who are considered creative, are given time to innovate and are consequently also expected to generate novel ideas and products. Other employees are in general not expected to innovate, they are not given time, and their ideas are not appreciated, not taken seriously, not understood, not accepted, not implemented and not rewarded. The question is: what would happen if these people were given the opportunity to innovate? Most of them would probably not know what to do or how to go about it. In general, people think that they don't know much about innovation and that they don't have the capacity to innovate – or perhaps they don't know that they actually have talents that could be used for innovation?

The question to be explored in this chapter is: what is needed for employees to drive and navigate innovation?

The structure of the text

The argument to be presented in the following is that project groups must nurture the development and practice of innovation competency and that organizations must develop leadership and create environments for supporting their employees in this endeavour. The main emphasis here will be on the first part: developing innovation competency. The study is mainly theoretical, but also draws on empirical data from an international executive master education, LAICS (Leadership and Innovation in Complex Systems), and from projects on developing innovation competency in corporations. Once the study has been positioned in relation to the contemporary societal conditions in general and to innovation in particular, innovation competency is defined, clarified and discussed. Then follows an argumentation that takes its point of departure from earlier research on innovation process by introducing the Diamond of Innovation (Darsø, 2001). The subsequent theoretical discussion aims at gradually building a theoretical framework for development of innovation competency in groups. The framework has been presented from a pedagogical perspective, that is, in relation to teaching innovation, in a recent Danish book (Darsø, 2011). In the present context, the framework will be introduced and argued from an organizational perspective, that is, related to developing innovation competency in the workplace. This is done by presenting the concept of learning space as an example of one approach, among others, to building innovation competency. This is discussed in relation to other theories, such as the knowledge spiral by Nonaka and Takeuchi. Whereas the main focus throughout the text is on micro-processes in groups, the last section focuses on the role of management and the organizational environment. This leads to a summary of insights and recommendations in the conclusion.

Innovation from different perspectives

Why is it important to develop innovation competency in organizations? The obvious answer is that in the twenty-first century, with accelerating change and high complexity, organizations must constantly renew themselves and their products – whether the products are material (such as food, clothes and medical equipment) or immaterial (such as services, software and experiences) (Pine and Gilmore, 1996). Globalization can be understood as both enabling and disenabling, as more customers and collaborators are within reach, but, of course, the same applies to the competitors. In innovation terminology, this recent openness towards exchange of ideas, new partnerships and strategic collaborations across organizational boundaries can be interpreted as open innovation (Chesbrough, 2005).

The concept of innovation, originally coined by Joseph Schumpeter in 1934, involves novelty that creates economic value (Schumpeter, 1934). Peter F. Drucker continued Schumpeter's work by elaborating and developing the concept of innovation towards practice (Drucker, 1985) and bringing it into a more modern context of the knowledge society (Drucker, 1993). From being mainly focused on technology and product development, innovation is now more and more being conceptualized as an integrated part of business and organizations. Innovation must form part of the strategy and business model (Kim and Mauborgne, 1999); it must be thought into the organizational culture (Yanow, 2000) and climate (Ekvall, 1991; Amabile *et al.*, 2002); it must enter IT systems and procedures (Mahnke *et al.*, 2006); it must be part of the leadership repertoire (Van de Ven *et al.*, 1999); it must influence project management (Cicmil *et al.*, 2009); it must be extended to involve users (Buur and Matthews, 2008; Thomke and Von Hippel, 2002); it must shape positioning and communication (Francis and Bessant, 2005); and, furthermore, organizations must have methods, toolboxes and processes that can advance and support all the phases of the innovation process (Darsø, 2001). The most important innovation component is, however, employees with knowledge, skills and competencies to create and develop innovative ideas and concepts that can go all the way to the client or the market. In the following, the specific focus will be on the development of innovative competencies of employees. This is not only important for the organizations; it is equally important for the career and future employability of the individual employee.

Innovation competency

The concept of competency can be defined as the ability to deal appropriately with situations as they arise. As Illeris points out, this kind of preparedness includes knowledge and social awareness combined with immediate judgement and decision-making (Illeris, 2011). Innovation competency is defined as the ability to create innovation by navigating together with others under complex conditions (Darsø, 2011). What does that entail? First of all, it entails a foundation of general knowledge about innovation. Being able to define what innovation is and what it's not is quite helpful – even though it's possible to innovate without this knowledge. Having basic knowledge such as who invented the concept and its historic origins, examples of innovation from one's own company as well as from other companies, together with knowledge about the innovation process and different types of process models can be considered part of this foundation. The basic part of innovation competency is thus knowledge *about* innovation, including having a shared vocabulary for understanding innovation. However, innovation competency can't be developed by studying theory alone; it must be developed through direct experience and practice. Certainly, theoretical frameworks, process tools, leadership roles and social technologies can be applied, but innovation competency emerges from dealing with tough challenges, difficult people and complex circumstances in real-life situations. Innovation competency can be understood as consisting of two intertwined competencies. The first is termed *socio-innovative competency*, which is defined as 'mastering social interaction that enhances innovation'

(Darsø, 2011: 176). This competency involves leadership, communication and facilitation. The second part is *intra-innovative competency*, which is defined as 'consciousness and sensitivity in relation to own and others' talents, preferences and potentials for development and innovation' (Darsø, 2011: 176). This competency involves inner transformation and growth towards a more sensitive, emphatic and conscious awareness of self and others. Metaphorically speaking, socio-innovative and intra-innovative competencies walk hand in hand and inspire each other. They are two sides of the same coin that is spinning the innovation process.

The Diamond of Innovation

In the following, I shall introduce a framework for developing innovation competency by explaining it piece by piece. Let me begin by introducing the Diamond of Innovation, developed in a three-year research project in a pharmaceutical company (Darsø, 2001). The research set out to examine the early innovation process, the conception and birth of innovation. The research question was: in what ways can innovative processes be initiated, supported and managed towards innovative crystallization in heterogeneous groups? One of the findings was that four parameters were central for initiating, supporting and managing the innovation process, and these could be displayed in a simple framework: the Diamond of Innovation (see Figure 6.1).

These four parameters are equally important for a successful innovation process. They can be comprehended and analysed as two dynamics: a knowledge dynamic between knowledge and ignorance, and a communication dynamic between relation and concepts. In most companies, the balance is skewed towards knowledge and concepts, that is, towards certainty

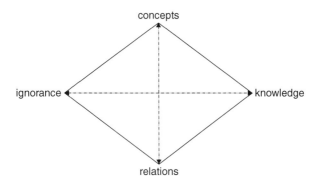

Figure 6.1 The Diamond of Innovation

and what can be seen and documented. This is understandable, because companies in general pay knowledge workers for applying their expertise and knowledge; but at the same time this is problematic, because, according to the research findings, the most innovative ideas were triggered in the area of uncertainty, that is, ignorance (ibid.). Surprisingly, knowledge often obstructed innovation. This was due to knowledge being a medley of data, opinions, attitudes, beliefs and unquestioned assumptions. Furthermore, it was found that relations seemed to be the decisive dimension for accessing the arena of ignorance, while concepts and conceptualizing were central for developing the innovative idea. In the following, we shall examine these dynamics more closely.

The knowledge dynamic: knowledge–ignorance

Knowledge has been pivotal in research and practice from around the time that Drucker coined the term 'knowledge worker' (Drucker, 1993). In the late 1990s, literature on the significance and use of knowledge in organizations boomed (e.g. Leonard-Barton, 1995; Nonaka and Takeuchi, 1995). Davenport and Prusak express this clearly in the introduction to their book *Working Knowledge* (Davenport and Prusak, 1998, p. xv):

> The core message of this book is that the only sustainable advantage a firm has comes from what it collectively knows, how efficiently it uses what it knows, and how readily it acquires and uses new knowledge.

Contrary to the general trend of the 1990s that knowledge could and should be contained in databases, the above-mentioned authors suggested that knowledge was embedded in people and that interaction was needed for knowledge to spread and grow. Nonaka and Takeuchi emphasized the dynamic process between tacit and explicit knowledge, forming a spiral process of organizational knowledge creation, and how this could lead to innovation. Hargadon has since written about brokering knowledge through *recombinant innovation*, which means combining existing knowledge in new ways, by moving knowledge and ideas from where they are known to areas where they are not known (Hargadon, 2002). Building on eight cases from consultant firms, Hargadon constructs a five-step model for linking learning and innovation:

1. *Access* to the often fragmented nature of different social worlds of organizations (project groups, work groups, departments)
2. *Bridging* strategies for exposing ideas and sharing knowledge
3. *Learning* about the activities people are involved in for contributing to working effectively and efficiently in the local environment
4. *Linking* by recognizing how knowledge from one context may apply in a different context
5. *Building* and constructing new networks around emerging innovations

Of special relevance here is the overall notion of recombinant innovation by linking what is known to what is not known. Recombinant innovation is in line with the dynamic between knowledge and ignorance in the Diamond of Innovation. In Hargadon's cases, the process of recombining knowledge is carried out by knowledge brokers, external or internal consultants, who help bridge problems from one small world with solutions from another. In the Diamond of Innovation framework, this process is managed by the participants through navigation tools that help exploit people's diverse perspectives in a project group. In relation to searching for novelty, the central idea is to explore, exchange and discuss what people know and, in particular, what they don't know. The enquiry towards knowledge takes the form of challenging what is known and of trying to dig under the surface to identify the underlying assumptions and challenge these. The enquiry towards ignorance takes the form of open questions of what, why, who and how. The idea is to encourage the participants to dive into what is not yet known, to try to see with new eyes and thereby to see new possibilities. This is not an easy process, as explained by the Canadian brain researcher Dr Bastiaan Heemsbergen:

> We can therefore be blinded by what we know. We can be blinded by our expertise, our mindsets, filters, paradigms, beliefs, orthodoxies, and rules of engagement, many of which we are not conscious or aware of. We are blind to what we don't know we don't know. (Heemsbergen, 2004, p. ix)

The suggestion that ignorance is an important ingredient of innovation is confirmed by Van de Ven and his colleagues (Van de Ven *et al.*, 1999: 81), who imply that 'the origination of true novelty should begin with profound ignorance.' In spite of this, the role of ignorance is still basically neglected in research on innovation.

The communication dynamic: relations–concepts

Relations and concepts are both ingredients of communication. Relations involve the intangible bonds between people, such as sympathy–antipathy, attraction–repulsion, inclusion–exclusion, trust–distrust. Relations are expressed in interaction, leadership processes and psychological climate. Concepts involve the more tangible content, such as the conversations, the ideas, the documentation, the minutes, the metaphors, the models: that is, the topical and cognitive content of communication. Both are involved simultaneously in communication and interaction. Relations form during conversations on topics of relevance to the project, but relations are tacit and often not part of the conscious process. A seminal study in 1964 by J.R. Gibb established trust to be the decisive factor for the quality of group climate and communication (Gibb, 1964). This has since been confirmed by Ekvall (1991) and Amabile *et al.* (2002). In relation to innovation, the findings from the research project on the early innovation processes suggest that the quality of relations set the foundation for groups daring to venture into the unknown – or not. The relations that encourage exploration and enquiry towards ignorance involve trust and respect (Darsø, 2001). As for concepts, this dimension entails all forms of conversations and can be enhanced by using metaphors, images and prototyping (Kelley and Littman, 2004; Schrage, 2000) as well as aesthetic and artful approaches (Darsø, 2004). Recently this dimension has been given more focus, in particular through current trends of cross-pollination between innovation and design research (Björgvinsson et al., 2010). All in all, communication and social interaction seem crucial for a successful innovation process, regarding both the quality of relationships and the depth and creativity of the conceptualization process.

The concept of preject

The Diamond of Innovation was created in order to provide a simple model for displaying the complexity of the early innovation process, which is difficult to grapple with when both of these dynamics are at work simultaneously. In order to differentiate this early phase from the later phase of a project, the term preject was coined. A preject covers the early (divergent) phase of *goal-seeking*, as distinct from the project, which concerns the later (convergent) phase with a *set goal* (Darsø, 2001: 196). Based on experience from practice, it seems, however, that preject and project are more useful concepts when conceived as two distinct modes of working, which ideally should be mastered by employees who work with innovation. Under normal circumstances, a preject would be followed by a project, but, because projects don't always evolve as expected, groups working with projects should be able to swap back and forth between the preject and the project mode of working. In explicit terms, the preject involves explorative search, which is directed by enquiry, whereas the project involves goal attainment, directed by tasks and milestones (Darsø and Høyrup, 2011). (See Figure 6.2.)

Innovation leadership roles

Since its conception, the Diamond of Innovation has been applied in various projects by a great variety of groups. Its main strength is that it can be used as a process map for navigating the innovation process. Another important feature is that it creates a shared language on innovation in organizations. Furthermore, four leadership roles, corresponding to the four parameters, have been developed in order to make the model more applicable and useful in practice (see Figure 6.3).

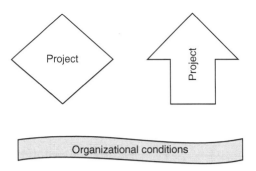

Figure 6.2 Preject–project

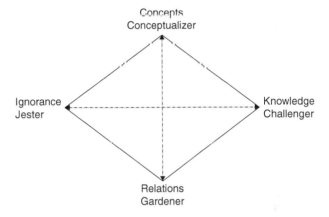

Figure 6.3 Leadership roles

Returning to the question of what is needed for employees to navigate the innovation process, the leadership roles (of gardener, jester, conceptualizer and challenger) are intended to help in driving the process. The idea is that, when using the Diamond of Innovation as an innovation process compass, the leadership roles are taken by four participants to perform the leadership needed for driving the innovation process during meetings. Ideally, the performance of the roles is discussed by all participants after each meeting in order to adjust and develop the leadership that this particular group finds appropriate and constructive. At the next meeting, other participants will take the roles and thereby show new qualities to be added. In this way, the desired leadership is gradually built around a learning process involving the development of socio-innovative competencies by creating awareness of what works, based on practical experience.

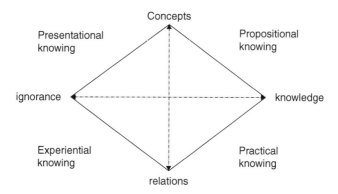

Figure 6.4 The Diamond of Innovation linked to four types of knowing

Adding four types of knowing

The Diamond framework with the leadership roles can be helpful for understanding the human and social aspects of innovation process, and thereby also the foundation for innovation competency. Applying the framework and practising the leadership roles gives direct experience, which gradually can turn into practical knowing and innovation competency. The process of building innovation competency can be understood as different ways of knowing. A framework that complements the dimensions of the Diamond of Innovation is the extended epistemology by Heron and Reason (2008). Their concepts of experiential, presentational, propositional and practical knowing are highly relevant for the learning processes that take place when working with the early phases of innovation. Innovation competency starts with experience and *experiential knowing* from being in the process, struggling with identifying powerful questions, trying to navigate in complex situations and contexts, and working on the social element of the innovation process regarding engagement, energy and commitment. The next step is to conceptualize the experience, the knowing and the learning, which can be done through *presentational knowing*. This concept concerns expressing the tacit or preverbal images[1] in presentational forms, typically artistic expressions such as pictures, drawings, prototypes or using music, dance or poetry (Damasio, 1999). From this, it becomes possible to express what has been learnt in verbal form as *propositional knowing*, such as sentences, statements, findings and conclusions. Knowing something with the mind, however, is far from *practical knowing*, which means embodying this knowledge. Getting to knowing in practice can, in fact, be a long journey, which involves several cycles of knowing.

In Figure 6.4, the four forms of knowing have been placed in the four arenas that are formed by connecting the four parameters of the Diamond. Experiential knowing is relevant in the arena between relations and ignorance because experiences are essential for getting people acquainted with each other as well as for creating relations of trust and respect. Furthermore, experiences are always new. Presentational knowing fits perfectly within the arena between ignorance and concepts because an important aspect of the innovation process is to identify what you don't know, and also to express what you don't know that you actually know. Conceptualizing vague images, ideas or hunches can be helped by artistic forms and by using prototyping techniques and materials. Propositional knowing obviously belongs to the arena between concepts and knowledge. In this arena, concepts are clear and can be stated and explained explicitly. Finally, practical knowing can be understood through the arena between knowledge and relations because this kind of knowing is tacit and mainly comes to the fore as behaviour embodied in practice.

A framework for developing innovation competency

The last layer of the framework concerns methods, tools and processes that support the learning cycle of the innovation process. Following the same sequence as earlier, in the first arena of experiential knowing the process that can support developing innovation competency is to create a *learning space*. This will be explained in more detail below. In the second arena of presentational knowing, the methods that can support developing innovation competency are *aesthetic and artful approaches* (Adler, 2006; Austin and Devin, 2003; Barry and Hansen, 2008; Darsø, 2004; Nissley, 2007; Taylor and Hansen, 2005; Taylor and Ladkin, 2009; VanGundy and Naiman, 2003). These methods work for clarifying and developing concepts, they serve as tools for reflection and communication and they can help elicit tacit knowledge. Aesthetic and artful approaches can help individuals and groups obtain new insights and are conducive for developing intra-innovative competencies (Darsø, 2011). In the third arena of propositional knowing, *innovation models and theories* can be applied as cognitive tools to think with, such as the innovation model by Van de Ven *et al.* (1999). In the fourth arena of practical knowing, the *technologies* that must be practised over and over are *social*. This concerns creating constructive innovation meetings as well as directing the explorative search by constructing questions that generate energy and commitment (Darsø, 2011; Vogt *et al.*, 2003). See Figure 6.5 with framework.

Creating a learning space

In order to understand how these methods can contribute to the development of innovation competence, we shall now focus on the notion of

118 Employee-Driven Innovation

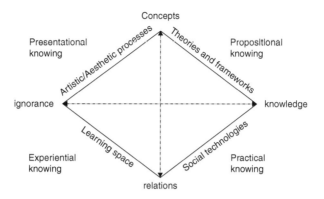

Figure 6.5 Framework for innovation competency

learning space. This will enable us to draw lines of comparison and extend the discussion from groups to departments or even to whole organizations regarding how organizational frames can support innovation. What is a learning space? A learning space is here defined as 'a psychological safety net, constructed by relations between people for creating broadness and community' (Darsø, 2011: 98). This definition is based partly on theory (Darsø, 2001) and partly on insights from the international master education mentioned in the introduction. The following extracts are from a master thesis:

> At the very first seminar in the education, it became very clear to me that this was going to be something I have never experienced before. The relationship-building was so intense and was a very profound part of the whole seminar, even though parts of this relationship-building was done as an integrated element in some of the other subjects we were presented to, so that a latent evolvement of bonding between us emerged almost subconsciously. And that has for me been *the* fundamental reason for obtaining a scene and a setting for optimal learning. (Stolt, 2008: 12)
>
> The social dynamics of our little group of four and the relations built up between us was a huge factor in both our work and in the reflection part. We had built up confidence, trust, and space for each individual and this scope made it possible for us to concentrate and contribute to the process in such a manner that everyone felt confident and secure. As discussed [earlier] the importance of relations is crucial. Actually I think it is *the* central element in an educational context. (Stolt, 2008: 13)

Illustrating how to create a learning space

Before drawing in other theories, further explanations are needed to throw light on how this type of learning space is created. In the following example, the emphasis will be on how we shape and build the learning space. This means that lectures and content discussions are omitted. Around 20 participants meet for the first time at a venue where they will spend three and a half days (three nights). After the first welcome, introduction of faculty and outlining the goals of the education (part-time, as all participants work in private and public organizations), and so on, we gather in the garden for an informal and playful presentation round. People are asked to create a 'body map' in relation to their place of work. This means that they will have to position themselves (their body) north, south, east or west of the other participants. In order to do this, they will obviously have to ask the others 'where' they stand, which means that they already begin to interact and create relations. When everybody has found their place, a faculty member walks around from person to person and invites them to say their name as well as a bit about themselves and their place of work. People will remember names and persons more clearly when relating memory to place, and in this way they also find out who works close to one another, which can be practical in many ways. A few more mappings are carried out, such as dividing the group according to working in public or private companies, and we often end the session by creating groups according to hobbies and interests.

The next step is to explain why we have done it this way. This involves a lecture on innovation, presenting the Diamond of Innovation, among others, and using that frame to explain the importance of creating relations of trust and respect. We explain that – apart from being fun – the mapping approach is, in fact, inspired by research. The participants are then asked to create a codex regarding how they want to be together in order to learn as much as possible. They are divided into small groups and given 15 minutes to clarify and discuss what is important in relation to feeling safe, acknowledged and included in the community. Each group contributes with a sentence or two to form the codex, which is written down and kept for future use and further amendments or changes. The codex process is a way of building trust. We end the session by asking for two volunteers to take on the leadership role of gardeners. As gardeners they will focus on the social atmosphere and be attentive towards the well-being of each participant, and they will also serve as the extended eyes and ears of the faculty for adjusting the program if needed. Each seminar will have two different gardeners in order for as many as possible to practise this leadership role.

At night, we have a storytelling session based on items that the participants have brought. They have each been asked to bring one item that tells something about who they are as individuals. Storytelling is eminent for creating interest and respect. As someone phrased it, 'you can't hate a person whose story you have heard.'

On the third day, a theatre director gives a master class on directing creativity through theatre rehearsal techniques. The purpose is twofold: first to comprehend creativity with both mind and body, and second to continue relation-building. The session, which involves theory, practical exercises and reflection, occupies morning and early afternoon and encompasses status games, communication exercises, bodily expressions as well as non-verbal exercises in pairs, in small groups and in the whole group. At the end of the day, people have worked on creative assignments, had eye contact, laughed together, touched each other and improvised; in short, they have had a lot of memorable experiences together.

To sum up in Heron and Reason's framework: during the first seminar, people have had numerous experiences together (experiential knowing), they have expressed thoughts and ideas in new ways, such as with their body (presentational knowing), they have been able to draw some explicit conclusions (propositional knowing), and they have practised and tried things out in practice (practical knowing).

Learning space and socialization

When we examine literature on concepts relating to learning space, the closest seems to be the notion of socialization by Nonaka and Takeuchi, the first step in their 'knowledge spiral' (Nonaka and Takeuchi, 1995). Their focus is on knowledge creation through knowledge conversion: 'the key to knowledge creation lies in the mobilization and conversion of tacit knowledge' (ibid.: 56). Socialization is the first of four steps in their model: Socialization (from tacit to tacit), Externalization (from tacit to explicit), Combination (from explicit to explicit) and Internalization (from explicit to tacit). *Socialization* concerns sharing experiences and building shared mental models through observation and direct experience. Typically this happens when people spend time together over a period of at least a few days, as, for example, when Honda would set up a brainstorming camp for discussing and solving difficult problems in development projects. 'Such a camp is not only a forum for creative dialogue but also a medium for sharing experience and enhancing mutual trust among participants' (ibid.: 63).

Learning space and 'Ba'

In a later article, Nonaka and Konno add the Japanese concept of 'Ba' to elaborate their model of knowledge creation. Ba means 'shared space for emerging relationships' and Nonaka and Konno propose to apply it for 'a shared space that serves as a foundation for knowledge creation' (Nonaka and Konno, 1998: 40). Four types of Ba are suggested to fit the four-step knowledge spiral: Originating Ba, Interacting Ba, Cyber Ba and Exercising Ba. Here, we shall focus on the first step: Originating Ba, which is linked to socialization. Originating Ba is created through face-to-face experiences and interaction, which at the same time enhance the creation of sympathy and empathy between people. 'From originating ba emerge care, love, trust, and commitment' (ibid.: 46).

Differentiation of learning space

There are obvious similarities between the above-mentioned concepts of Nonaka and his colleagues and the concept of learning space. Key words are face-to-face experiences, trust building, dialogue and community. Nonaka and Takeuchi examine knowledge creation, both the combination of existing knowledge (similar to Hargadon's recombinant innovation) and the creation of new knowledge. The main difference is the degree of process facilitation. How much and in what way should this process be directed? Apparently, in Nonaka and Takeuchi's set-up, socializing happens without intervention. In our educational setting, this process is directed and facilitated, but not controlled. Facilitation comes from Latin and means making things easy, and this is done through frame-setting, invitation and structure. Based on many years of practice, I will argue that well-contemplated frames are generative of learning, creation and innovation.

In the LAICS education, the socializing process is encouraged and made explicit by making people conscious about its significance and the effects gained by participating. The purpose is thus twofold: first, to create the learning space for optimizing learning throughout all the seminars; second, by having experienced how it is done and what it means, the participants can learn how to create a learning space for others, such as in their project groups or in their workplace. This forms part of socio-innovative competency development.

Learning space in organizations

Going back to the organization, we can identify parallels from the concept of learning space to setting the organizational frames for developing innovation competency. This will be the focus of the final section before reaching the conclusion. Frames in general can be physical, mental, social, emotional and structural. Evidently, management and leadership as well as organizational culture influence framing. Here we must necessarily narrow it down to structural and social frames, including leadership, and to the more accessible layer of culture: climate.

Climate

Climate is defined by Professor Göran Ekvall as the behaviour, attitudes and feelings which are fairly easily observed (Ekvall, 1991: 74). Ekvall sees climate as an intervening variable, which influences communication, problem-solving and decision-making as well as motivation and learning (ibid.). Climate also influences operations and outcomes. In his ten-year research programme, Ekvall isolated nine climate dimensions: challenge, freedom, idea-support, trust, dynamism, playfulness, debates, conflicts, and risk-taking. Innovative organizations score high on all these dimensions except conflict. Stagnating organizations score low on all dimensions except conflict.

Intrinsic motivation

Harvard professor Teresa Amabile, who has carried out numerous research projects on how to foster innovative workplaces, concludes that the component that can be influenced most easily by the work environment, and is also the most influential, is *intrinsic motivation*. Amabile strongly recommends that managers should establish a work environment with the right amount of challenges and freedom, in line with Ekvall's findings. In order to do that, managers must care about, understand and support people in their workgroups and people must learn to help and respect each other in their groups. Amabile found people to be most creative – even under stress – when they feel that their work matters. And it is the task of the manager to show keen interest and to protect and support such workgroups. Unfortunately, what happens more often in organizations is that 'creativity gets killed' (Amabile, 1998).

Creative leadership

Similar findings can be seen in studies of theatre directors' work with ensemble rehearsal. The most inspiring theatre directors work closely with their actors:

> They possessed a powerful aura of focused concentration. As an actor, it was this sense of being rigorously observed that made them so stimulating to work for... They would not fail to notice when you succeeded in something and you had a sense they were merely indifferent to your failures. There was an expectation implicit in the behaviour of these directors, they were looking for something rare, they knew it might not be easy to find... When they spotted it, they supported and encouraged. When you were searching, they let you wander about, but when you were stuck, no longer flowing with ideas, they would assist, either with specific technical input or with creative challenges – What if?-questions.
> (Ibbotson and Darsø, 2008: 554)

Theatre ensembles need to build safe learning spaces in order to create and improvise newness, just as innovative teams need to build relations of trust and respect in order to sustain uncertainty and not knowing where the process will take them. The kind of direction found in theatre rehearsal could inspire and spark new kinds of leadership in organizations. More than half a century ago, Peter Drucker advocated for management by walking around. In many organizations of today, we find management by email in a secluded office. Being present, showing interest and caring for employees and their projects, as Amabile suggests, would constitute a supportive environment for encouraging innovation and the development of innovation competency.

Conclusion

In this chapter, I have argued for developing innovation competency in organizations. Based on earlier research on innovative groups, the Diamond of Innovation was suggested as a constructive framework for making sense of complex social innovation processes. The model serves as a compass for understanding and navigating explorative search enquiries necessary for identifying and developing innovative concepts. In order to support navigation, four leadership roles were presented for driving the innovation process successfully. Innovation competency can be understood as the ability to create innovation by navigating in complex processes together with others. The competency builds on a foundation of knowledge about innovation complemented with experience and practice in innovation work. Just as artists need to practise their instrument daily, people working with innovation need to practise enquiring, searching, navigating, interacting and driving the innovation process as often as possible. Only through daily practice – and through welcoming both failures and successes – will people be able to develop the innovation competency that is highly needed in global society. Some of the social competencies surely exist in organizations already, but need to be recognized and put to use.

Adding Heron and Reason's four ways of knowing to the Diamond framework complemented and clarified the learning process by emphasizing experience and practice. The Diamond of Innovation framework was completed by adding four types of methods and tools in order to explain how innovation competency can be encouraged in practice. Unfolding and illustrating the creation of a learning space was chosen because this concept could most easily be extended to an organizational context. Indeed, several similarities were identified between learning space and Ekvall's research on the organizational climate characteristic of innovative organizations, and, similarly, we found confirmation in Amabile's extensive research on the relation between intrinsic motivation and creativity in organizations. Amabile has also examined the role of management in designing environments that encourage intrinsic motivation and innovation. For innovation competency to develop, evidently a matching leadership development is needed. Ideally, managers should themselves have gone through some kind of practice in innovation. This would also provide the advantage of building a shared language for innovation, which is lacking in many organizations, but helpful for communication. In an ideal world, managers would develop innovation competency themselves in order to understand and nurture it in their organizations.

Establishing learning spaces in groups and organizations would support healthy and creative work environments. In a real world, one of the main challenges would be to alter an already established culture. Other things being equal, it is easier to create a learning space with a group of people who don't know each other (as in the LAICS education), and who have consequently not yet formed relationships, than to attempt to change an existing working practice. But building learning spaces in relation to innovation projects would be a good start – especially if supported by management.

Note

1. Damasio (1999, p. 9): 'By *image* I mean a mental pattern in any of the sensory modalities, for example, a sound image, a tactile image, the image of a state of well-being.'

References

Adler, N.J. (2006). 'The arts and leadership: now that we can do anything, what will we do'. *Academy of Management Learning and Education*, 5(4), 486–499.
Amabile, T.M. (1998). 'How to kill creativity'. *Harvard Business Review* (HBR Collection: Best of HBR on Rescuing Your Company's Creativity) 76(5): 76–89.
Amabile, T.M., Hadley, C.N. and Kramer, S.J. (2002). 'Creativity under the gun'. *Harvard Business Review* (HBR Collection: Best of HBR on Rescuing Your Company's Creativity) 80(8): 52–61.
Austin, R. and Devin, L. (2003). *Artful Making. What Managers Need to Know about How Artists Work*. New Jersey: FT Prentice Hall.
Barry, D. and Hansen, H. E. (2008). *SAGE Handbook on New Approaches in Management and Organization*. London, Thousand Oaks, New Delhi and Singapore: Sage Publications.
Björgvinsson, E., Ehn, P. and Hillgren, P.-A. (2010). 'Participatory design and "democratizing innovation"'. Paper presented at the Participatory Design Conference, Malmö.
Brunson, N. (2003). 'Organized hypocrisy'. In Czarniawska, B.G.S. (Ed.) *The Northern Lights: Organisation Theory in Scandinavia* (pp. 201–215). Oslo: Copenhagen Business School Press.
Buur, J. and Matthews, B. (2008). 'Participatory innovation'. *International Journal of Innovation Management*, 12(3), 255–273.
Chesbrough, H.W. (2005). *Open Innovation. The New Imperative for Creating and Profiting from Technology*. Oxford: Oxford University Press.
Cicmil, S., Hodgson, D., Lindgren, M. and Packendorff, J. (2009). 'Project management behind the façade'. *Ephemera*, 9(2), 78–92.
Damasio, A. (1999). *The Feeling of What Happens. Body and Emotion in the Making of Consciousness*. San Diego, New York, London: Harcourt, Inc.
Darsø, L. (2001). *Innovation in the Making*. Frederiksberg, Denmark: Samfundslitteratur.

Darsø, L. (2004). *Artful Creation. Learning-Tales of Arts-in-Business*. Frederiksberg, Denmark: Samfundslitteratur.

Darsø, L. (2011). *Innovationspædagogik. Kunsten at fremelske Innovationskompetence*. Frederiksberg: Samfundslitteratur.

Darsø, L. and Høyrup, S. (2011). 'Developing a framework for innovation and learning in the workplace'. In Melkas, H. and Harmaakorpi, V. (Eds) *Practise-Based Innovation: Insights, Applications and Policy Implications*. Berlin Heidelberg: Springer-Verlag.

Davenport, T.H. and Prusak, L. (1998). *Working Knowledge. How Organizations Manage What They Know*. Boston, Massachusetts: Harvard Business School Press.

Drucker, P.F. (1985). *Innovation and Entrepreneurship: Practice and Principles*. London: Pan Books.

Drucker, P.F. (1993). *Post-capitalist Society*. Oxford: Butterworth-Heinemann.

Ekvall, G. (1991). 'The organizational culture of idea-management: a creative climate for the management of ideas'. In Henry, J. and Walker, D. (Eds) *Managing Innovation* (pp. 73–79). London, Thousand Oaks, New Delhi: Sage Publications.

Francis, D. and Bessant, J. (2005). 'The four Ps: targeting innovation and implications for capability development'. *Technovation*, 25(3), 171–183.

Gibb, J.R. (1964). 'Defensive communication'. *Journal of Communication*, 11, 141–148.

Hargadon, A.B. (2002). 'Brokering knowledge: linking learning and innovation'. *Research in Organizational Behavior*, 24, 41–85.

Heemsbergen, B. (2004). *The Leader's Brain: How are You Using the Other 95%?* Victoria, Canada: Trafford.

Heron, J. and Reason, P. (2008). 'Extended epistemology within a co-operative inquiry'. In Reason, P. and Bradbury, H. (Eds) *The SAGE Handbook of Action Research Participative Inquiry and Practice* (pp. 366–380). London: Sage Publications.

Ibbotson, P. and Darsø, L. (2008). 'Directing creativity: the art and craft of creative leadership'. In Kerr, C. and Darsø, L. (Eds) *Journal of Management and Organization, Special Issue: Re-conceiving the Artful in Management Development and Education*, 14(5), 548–559.

Illeris, K. (2011). *Kompetence. Hvad Hvorfor Hvordan?* Frederiksberg: Samfundslitteratur.

Kelley, T. and Littman, J. (2004). *The Art of Innovation*. London: Profile Books.

Kim, W.C. and Mauborgne, R. (1999). 'Creating new market space: a systematic approach to value innovation can help companies break free from the competitive pack'. *Harvard Business Review* (January–February), 83–93.

Leonard-Barton, D. (1995). *Wellsprings of Knowledge. Building and Sustaining the Sources of Innovation*. Boston, Massachusetts: Harvard Business School Press.

Levitt, T. (2002). 'Creativity is not enough'. *Harvard Business Review*, 80(8), 137–145.

Mahnke, V., Özcan, S. and Overby, M.L. (2006). 'Outsourcing innovative capabilities for it-enabled services'. *Industry and Innovation*, 13(2), 189–207.

Nissley, N. (2007). 'Framing arts-based learning as an intersectional innovation in continuing management education: the intersection of arts and business and the innovation of arts-based learning'. In DeFillippi, B. and Wanke, C. (Eds) *University and Corporate Innovations in Lifetime Learning*. Greenwich, Connecticut: Information Age Publishing.

Nonaka, I. and Konno, N. (1998). 'The concept of "ba": building a foundation for knowledge creation'. *California Management Review*, 40(3), 40–54.

Nonaka, I. and Takeuchi, H. (1995). *The Knowledge-Creating Company: How Japanese Companies Create the Dynamics of Innovation*. New York: Oxford University Press.

Pine, B.J.I. and Gilmore, J.H. (1996). *The Experience Economy. Work is Theatre and Every Business a Stage*. Boston: Harvard Business School Press.

Schrage, M. (2000). *Serious Play. How the World's Best Companies Simulate to Innovate.* Boston, Massachusetts: Harvard Business School Press.

Schumpeter, J.A. (1934). *The Theory of Economic Development.* Cambridge, Massachusetts: Harvard University Press.

Stolt, J. (2008). *New Ways of Learning in Entrepreneurship Education.* Copenhagen: Aarhus University.

Taylor, S.S. and Hansen, H. (2005). 'Finding form: looking at the field of organizational aesthetics'. *Journal of Management Studies*, 42(6), 1211–1232.

Taylor, S.S. and Ladkin, D. (2009). 'Understanding arts-based methods in managerial development'. *Academy of Management, Learning and Education*, 8(1), 55–69.

Thomke, S., and Von Hippel, E. (2002). 'Customers as innovators. a new way to create value'. *Harvard Business Review* (April 2002), 74–81.

Van de Ven, A.H., Polley, D.E., Garud, R. and Venkataraman, S. (1999). *The Innovation Journey.* Oxford, New York: Oxford University Press.

VanGundy, A.B. and Naiman, L. (2003). *Orchestrating Collaboration at Work: Using Music, Improv, Storytelling, and other Arts to Improve Teamwork.* San Francisco: Jossey-Bass/Pfeiffer.

Vogt, E.E., Brown, J. and Isaacs, D. (2003). *The Art of Powerful Questions. Catalyzing Insight, Innovation, and Action.* Mill Valley, California: Whole Systems Associates.

Yanow, D. (2000). 'Seeing organizational learning: A "cultural" view'. *Organization: The Interdisciplinary Journal of Organization, Theory and Society*, 7(2), 247–268.

7
Employee-Driven Innovation and Practice-Based Learning in Organizational Cultures

Ulrik Brandi and Cathrine Hasse

How can we understand employees as drivers of innovation and what prevents employees from being realized as an innovative capacity in organizations? The present definition of EDI contains a discrepancy: on the one hand, EDI is claimed to cover purely bottom-up processes, while, on the other hand, empirical examples show that EDI is dependent on a cultural context in which the employees' everyday creative actions (based on practice-based learning) are recognized as potential resources for innovation in the organization. Due to the lack of a concept of organizational culture in relation to the analysis of employee-driven innovation, we are unable to grasp fully why attempts to be innovative sometimes fail, and, which is even more widespread, why practice-based potential innovation is never realized. In this chapter, we will thus improve our knowledge of why innovation must be recognized as practice-based in the organizational culture.

As Fagerberg and Verspagen (2009) demonstrate, the research community of innovation is a multidisciplinary field with a long-standing research tradition. Schumpeter, with his seminal work on economic development, marks the beginning of this vibrant and voluminous research field, and in 1928 he defines innovation as follows: 'What we, unscientifically, call economic progress means essentially putting productive resources to uses hitherto untried in practice, and withdrawing them from the uses they have served so far' (Schumpeter 1928: 378). This is what we call 'innovation'.

We follow the Schumpeterian definition in general terms and understand innovation as a qualitative, new combination of existing resources, experiences and knowledge aiming at generating improvement and novelty in either processes or products. On the basis of this definition, we ask how new combinations of existing resources and knowledge executed by employees on the basis of practice-based learning processes are recognized as innovation by the organization. Putting employees' productive resources to uses hitherto untried in practice requires a supportive organizational culture.

To contemplate these considerations in an organizational setting, we discuss in this article the concept of EDI in relation to several empirical examples, and we present a typology for how organizational cultures relate to the innovation potential of practice-based EDI. Thus, in this article we intend to enquire more deeply as to how we can understand the concept of EDI, and how EDI and organizational culture are entangled.

Innovation driven by employees

The Schumpeterian seminal work and ideas have been developed over the past 25 years in neo-Schumpeterian economics, with emphasis on the importance of innovation as the driver of economic and societal dynamics and development (Hanusch, 2007; Winter, 2006). Scholars and politicians all stress the importance of facilitating innovative competencies in workplaces as the primary parameter for economic and social progress and success in modern societies and organizations (Amabile, 1988: 1740; Arundel, 2007; Reich, 1991: 33; Wolfe, 1994). Arguing that innovation is integral to economic success in organizations and societies naturally draws attention to one main resource that gives life to the strong normative claims of these statements: the employees' innovative abilities in the workplace. We thus see a broad interest among scholars and politicians in attempting to explain how innovative abilities in the workplace can be supported and implemented in creating new products or processes.

In the literature of the field, we find different discourses on how to conceptualize innovation. These discourses view the employees as constituting an innovative force in both a direct and a more indirect way. In the more indirect approach, the researchers and politicians are mainly interested in understanding how workplaces and governmental structures can support the development of innovative competencies in the workforce, thus increasing the general level of human capital in the workplaces. Hence, in this approach the research question is how we develop the innovative competencies of the employee. In the more direct approach, the employee is assumed to possess innovative competencies, so that the interest is instead directed at how the employee functions as a driver for innovation in her/ his workplace. Here, the research question is how innovation is driven by the employee and what facilitates and impedes this type of innovation process. Our starting point is the latter research question. In this more direct approach to innovation driven by employees, we see three general theoretical understandings of how to conceptualize innovation driven by employees.

First, there is what we call the R&D approach. Here, managers and employees in R&D departments must be innovative *per se*, whereas other employees are not expected to take part in innovation processes. Godin (2006: 639) terms this the linear model of innovation, where the linearity is characterized by progression from basic to applied research, leading to development of research-based knowledge and ending with production and diffusion of the knowledge internally or externally. In the R&D approach, the employees' innovative work is strongly formalized and organized in the workplace.

The second understanding of innovation driven by employees underlines the necessity for all employees to be innovative and is termed 'corporate entrepreneurship' (Åmo, 2006a; 2006b). Åmo writes about employee innovation behaviour as something that must be supported by corporate entrepreneurship programmes, where management makes it clear to employees that innovative behaviour is desired (ibid.). Corporate entrepreneurship sees economic growth as something that originates from the employees' innovation initiatives and as dependent on the organization's intellectual capital (Floyd, 1999; Hayton, 2005). We find a related concept of employee-driven innovation at the workplace in the work of Janssen (2000), who talks about innovative work behaviour and relates innovation to the individual employee.

Again the basic premise of why we should be interested in innovation driven by individuals at the workplace is that it can ensure long-term survival and growth of organizations (ibid.: 287). Here, too, it is the individual capacity to innovative work behaviour that is investigated and measured. The traditional understanding measures the employees' innovation by the number of suggested ideas or changes in the workplace. The continuous improvement is measured by the number of ideas put forth in employee suggestion schemes (see also Bonnafous-Boucher *et al.*, this volume). Wynder (2008: 356), for instance, enquires into the continuous improvement of workplaces as a kind of employee-driven innovation important for organizations' economic growth. The relation between management and employees becomes crucial when it is the responsibility of management to sort bad ideas from ideas that are worth pursuing (ibid.).

A third strong discourse and understanding of innovation driven by employees is found in the work of Amabile *et al.* (1996), Amabile (1988) and Oldham and Cummings (1996). This discourse is characterized by a major interest in the creativity of the individual, often understood as the creation and suggestion of ideas. In this discourse, all employees can come up with creative ideas, but the individual creative act is perceived to be separate from the realization of the idea. Creativity is seen as the production of novel ideas in all aspects of work life, whereas innovation is defined in relation to the organization's ability to make use of and successfully implement the creative ideas (e.g. Amabile *et al.*, 1996). The relevance of this discourse has been emphasized in later works by, for example, Gupta (2007: 886), who argues for two different strands of individual creativity: first, an understanding that explains creativity as a personal trait, while the second strand is an understanding of individual creativity as closely related to the social context of the workplace. In this third discourse on employees as an innovative capacity, Axtell (2000: 266) argues for discriminating between factors that influence suggestion of ideas and implementation of ideas. One important distinction in the analysis of Axtell is that suggestion of ideas is primarily related to personal traits, whereas idea implementation is highly determined by organizational factors.

Summing up, this short general review of the field of innovation driven by employees presents two important issues. First, we see a separation between literature that has a focus on R&D personnel and literature concentrating on all employees as contributors of innovation. Put differently, in parts of the literature concerning innovation only especially appointed R&D personnel are expected to be innovative, whereas in other parts all employees are seen as being able to contribute to the corporate entrepreneurship by providing new ideas. The common focus is, however, on the innovative/creative abilities of individual employees in organizations.

Second, the focus in parts of the literature on the separation between individual employees' creative ideas and the implementation of these ideas in the organization raises new questions about the relation between management, the individual creative employee and the organizational culture in which the ideas are materialized and implemented.

Managing employee-driven innovation: content and challenges

The concept of employee-driven innovation was, in a Danish context, initially a policy concept coined by the Danish trade union LO, The Danish Confederation of Trade Unions. In 2005, LO published a report on employee-driven innovation in public and private enterprises (conducted by the consultancy agency Rambøll Management). The report presented nine cases of public and private enterprises which were engaged in employee-driven innovation. In all cases, the management supported the participation of employees in employee-driven innovation.

In this report the concept of employee-driven innovation was defined as follows: 'Employee-driven innovation refers to situations where employees, in a broad sense, contribute actively and systematically to the innovation process' (Rambøll Management 2006: 1, authors' translation). Another definition also includes management, when it is described as 'the will and ability of management and workers to take initiatives, to improve the quality of goods and services, to make innovations and to develop the production process and consumer relations' (European Foundation for the Improvement of Living and Working Conditions, 1997: 15).

The second approach shares a focus on employee innovation as being somehow connected to management decisions. This view is supported in the general literature on involvement of employees in developments of innovation; they are often sanctioned and initiated by management, but this is not what is meant by employee-driven innovation (Høyrup 2010: 148). Employee-driven innovation is defined as a bottom-up process including 'innovative practices, contributed by any employee (outside the boundaries of his/her primary job responsibilities), at all levels of the organization' (Høyrup 2010: 152). So far, the focus in EDI research has been on employee-driven innovation as a bottom-up process, which can emerge spontaneously and informally or be 'organized and supported by various organizational and managerial means' (Høyrup, 2010), but it has not been discussed how employee-driven innovation is *recognized* by management, when it occurs spontaneously, or how the employees' practice-based learning is taken into account and supported by management in the everyday workplace culture.

In all of the cases reported by Rambøll[1]– and indeed in the many initiatives initiated by the Danish state and others – the motivation for initiating employee-driven innovation comes from outside the employees' everyday practice-based learning. Yet, to make employee-driven innovation a more sustainable concept to work with, we have to delve deeper into the aspects of how employee-driven innovation can be supported, when it is not initially encouraged by management or state funding.

What we shall add to the existing definitions of employee-driven innovation is that the innovation is particularly valuable for organizations because we can characterize it as *practice-based learning*. With reference to Per-Erik Ellström's work, innovation that *emerges from employees'* practice-based learning has been defined as follows: '[The] innovation process of implementing new methods, working procedures, routines and services is based on the experience, knowledge and skills which employees have acquired by engaging in the working processes of the firm' (Høyrup, 2010: 151) in which the basic mechanism is learning. Yet, the very concept of learning often remains a black box.

We wish to know more about what invites or closes processes of learning in order to show how an organizational culture can come to recognize employees as drivers of innovation. We wish to study the relation between intended acts of innovation and the implementation and learning of organizational innovation in local workplace cultures which are open and closed to the recognition of innovative capacities. In this clash we see the relevance of bringing in a new concept of organizational culture associated with learning to improve our understanding of how EDI is realized.

Pairing innovation and culture

To study the relation between innovation and organizational culture is not a novel theme in innovation research. The vast number of publications on innovation and culture viewed from group to country level and from product to process is clear evidence of this statement (Ahmed, 1998; Gudeman, 1992; Herbig and Dunphy, 1998; Lemon and Sahota, 2004; Morcillo *et al.*, 2007). A review of the majority of the core texts in this field demonstrates the primacy of one type of relation between innovation and organizational culture. This type of relation understands culture as an important factor influencing the innovation process and outcome (see, e.g., Ahmed, 1998; Chiu and Liaw, 2009; Feldman, 1988; Kiurunen, 2009; Lin, 2009; Martins and Terblanche, 2003; McLaughlin *et al.*, 2008; Tan *et al.*, 2008). From this perspective, innovation is dependent on specific cultural traits as factors that can either facilitate or impede innovation. Organizations can *have* organizational culture in a top-down instrumental and applicable way (Smircich, 1983). Hence, the direction of this type of relation between innovation and organizational culture goes from culture to innovation (organizational culture → innovation).

However, the direction of this relation can also be the other way around, so that an innovation is not an innovation until it has been recognized as an innovation in a specific cultural context. What is recognized as an innovation is related to cultural–historical, practice-based learning processes. Thus, when practice-based learning differs *within* an organization, different workplace cultures occur within the organization, and these may or may not be recognized by the management, which we argue is embedded in its *own* workplace culture. Consequently, an organizational culture can consist of a number of different local practice-based workplace cultures. What is innovative in *one* local workplace culture might not be recognized as innovative in the wider organizational culture. This is a different way of conceptualizing the relation between innovation and organizational culture than the one we primarily find in the research literature on innovation. The direction of this type of bottom-up relation goes from innovation to culture (innovation → organizational culture). We see this as an often neglected approach in innovation studies. This approach puts emphasis on the importance of learned collective, cultural patterns of meaning and values for the recognition and enactment of innovation.

As already mentioned, the relation between the fields of innovation and organizational culture is established. However, as suggested earlier, this relation has not, to any appreciable extent, taken into account development within the field of organizational culture. Beginning with an explosive interest in the 1980s, research, organization and management studies have made a particular darling of the concept of organizational culture; *Journal of Management Studies* (1982, 1986), *Organisational Dynamics* (1983), *Administrative Science Quarterly* (1983), *Journal of Management* (1985), *Organization Studies* (1985) and *International Studies of Management and Organization* (1987) all launched special issues on this new field, and a number of important books spurred interest in this new promising field (Parker, 2000: 59).

What characterized books like Terrence Deal and Allan Kennedy's *Corporate Cultures* from 1982 was a strong belief in a management-driven culture; that is, culture was seen as an entity that could be managed and controlled by leadership. In this 'objectivist perspective' promoted by many studies in organizational culture to this day, cultural orientation is considered 'one of the organizational variables' (Mavondo and Rodrigo, 2001: 245). Definitions of organizational culture by, for example, Edgar Schein (1992) and Deal and Kennedy (1982) have been accused of being academically superficial consultant-like definitions (Alvesson and Berg, 1992: 33), which miss the complexity of everyday life in organizations. Empirical studies have made it clear that, whatever culture is, it is not easily manageable. Management does not create culture; members do (Martin, 2002). In the wake of massive interest in how to boost innovation in organizations, it seemed logical to many that innovation studies would see culture as yet another tool to improve and enhance innovation. The promise of an 'innovation culture' is built into a good number of arguments. Many, for instance Mavondo and Farrell (2003), are aware that the idea of culture as a variable that is easy to manipulate must be repudiated. Nevertheless, they choose to maintain that '[m]ost marketing researchers treat culture as something the organization has and have demonstrated that market oriented cultures enhance organizational performance', that '[c]ultural consistency creates economies of horizontal and vertical coordination since subordinates know how their boss would like things to be done' and that culture is a strategic asset as it is 'significantly and positively related to innovation' (ibid.: 241–242).

The concept of culture has gone from a description of what an organization *has* (which can be manipulated in line with other variables) to something it *is* (Smircich, 1983) in a clearer understanding of what the concept of culture covers and how culture, rather than management alone, drives organizations. Researchers adopting this approach will explore cultural manifestations in the shape of, for example, symbols and search for patterns of meaning behind behaviour through 'thick descriptions' of everyday life (Geertz, 1973). Their aim is to 'gain an in-depth understanding of how people interpret these manifestations, and how these interpretations form patterns of clarity, inconsistency and ambiguity that can be used to characterize understandings of working lives' (Martin, 2002: 4–5).

Conceptualizing organizational culture and practice-based learning

A closer look at the complexities of everyday life in organizations might even gradually dissolve the very notion of a shared culture – initially into subcultures and finally into a fragmented perspective without any clear presence of anything shared (Martin, 1992). This closer look at the complexities has ultimately resulted in 'deconstructing the idea of collective culture altogether' (Hatch and Schultz, 2004: 338).

The movement from culture as something an organization has to something it *is*, and finally to something consultants and researchers *write* about and can be deconstructed, has, however, not meant an end to the interest in organizational culture. On the contrary, as Carlos Gonzales reminds us: 'Even though the problem of culture has not been solved, in recent years there has been an explosion of cultural inquiry within international management' (Gonzales, 2008: 95).

There is a risk that glossing over the critique of the culture concept in relation to innovation could lead to a repetition of the culture 'bubble' of the 1990s, which ended in accusations of organizational cultural analyses promising more than they could fulfil because, at the end of the day, cultural analysis turned out to be subjective, imprecise, one-sided and unrecognizable for ordinary employees in the everyday life of the organizations (Alvesson and Berg, 1992; Martin, 1992; Parker, 2000; Smircich, 1983).

Instead of ignoring the critique, we propose new efforts to improve and broaden the culture concept to explain not only cases where an organizational culture leads to a boost in innovation capacities (e.g. Lemon and Sahota, 2004) but also cases where attempted innovation fails in the meeting with the given organizational culture.

In our definition, organizational culture builds on the various local workplace cultures that are created through subtle and complex processes of practice-based learning. Organizational culture is in constant development and simultaneously affected by and contributing to a collective activity at the workplace, which leads certain workplace cultures either to exclude some values, ideas and members from participating in this activity or to include or support their formation as part of the culture. Put differently, we define organizational culture as what we *do* in organizations, which teach employees certain cultural values, discourses, emotions, traditions and meanings of artefacts through a process of practice-based learning. This entanglement of materiality and discourse largely forms shared self-evident and taken-for-granted patterns of meaning underlying the inclusion and exclusion of ideas and people in workplace cultures (Hasse, 2011; Hasse and Trentemøller, 2011).

This definition of a workplace culture is different from many other definitions in which the focus has been on consensus – most notably the definitions developed by Edgar H. Schein and his followers. Here, consensus is obtained by shared learning, which covers behaviour, emotional and cognitive elements of the total psychological function of the group members. A particular workplace culture can make it more or less difficult or rewarding for certain types of employees (for example low-skilled employees or women) to show their creative potential. This was especially noted in a previous study of gender-segregated workplace cultures (Hasse and Trentemøller, 2008).

Even if these marginalized employees do take initiatives, the initiatives may not be recognized as creative and innovative – or may even be perceived as elements of irritation (ibid.: 178) – because the employees' actions are not recognized in the wider context of the institutional activity by the management or local leaders.

Instead of a traditional focus on organizational learning as related only to 'correcting and solving problems', we suggest that organizational learning is understood as the learning that arises from the everyday chores of the practice and brings a group of people together in social designation (Hasse, 2008). In this sense, an organizational culture constitutes a community of practice in which a newcomer gradually learns to become an old-timer in the practice, which holds the community together (Lave and Wenger, 1991: 57). In this process, newcomers also learn the specific cultural values, discourses, emotions, traditions and meanings of artefacts already learned by the old-timers, and the organizational culture is thus linked to the process of practice-based learning (e.g. Brown, Collins and Duguid, 1989; Gherardi, 2000; Lave and Wenger, 1991; Nicolini *et al.*, 2003). When employees of one overall organization, like a municipality, learn, their practice-based learning is guided by motives linked to local values, discourses, emotions, traditions and meanings of artefacts (e.g. Engeström, 2001). The challenge for management is how to recognize local practice-based cultures as respected activities in relation to the overall organizational culture recognized by the management.

One way to tie local practice-based cultures together in an inclusive organizational culture is through management's attention to the importance of recognizing the values linked to the employees' work life when they come up with creative ideas. EDI might be an important form of everyday practice-based learning (Fenwick, 2003: 123), but is EDI recognized by those in power as a potential innovation that can be realized in the overall organization? We argue that innovations in organizational cultures are tied to what people do as a community of practice, and management cannot take for granted that their understanding of innovation is based on the presumed shared basic assumptions, discourses, values, artefacts, emotions, traditions and meaning-making of the overall organization.

Illustration of three EDI cases

The cases we analyse illustrate how different workplace cultures open up for the managers' recognition or rejection of how the employees, who are engaged in practice-based activities, can innovate in the organization. In the first case, the managers act as drivers of an attempted innovation that fails because it is not in accordance with what the employees can recognize as an innovation from their practice-based perspective. In the second case, an employee informally learns how to solve a problem with changing a patient's bed sheet by putting productive resources of garbage bags to uses hitherto untried in practice. However, the management never hears of this potential innovation and it thus remains a creative action that does not influence the general workplace culture. In the third case, an employee gets a good idea, which is not recognized in the organizational workplace culture but leads to a new organization, with a culture more open to employee-driven innovation. Together, these three very different cases set the stage for a typology for recognition of bottom-up employee-driven innovation in organizations.

Case 1: 'the new office'

In the first case, the empirical data comes from a research project conducted in a large Danish municipality north of Copenhagen. This study is the only case we present which builds on an extensive fieldwork lasting several months followed by a number of interviews. The research focused on the service area for children with special needs, which includes children with behavioural, physical and psychological problems. In Denmark the public service area for children with special needs is traditionally organized into two types of administration, each engaged in the local practice-based learning experiences of the local workplace cultures and their respective values:

1) Social administration (SOADM): administration of social affairs from social service centres and institutions that are responsible for solving problematic cases of a social character involving not only the children but the entire family (typically parents and siblings) in the problem-solving activity.
2) Educational administration (EDUADM): administration of educational affairs from daycare institutions, schools and pedagogical–psychological service departments that are responsible for solving cases of an educational and psychological character for the individual children.

On 1 January 2005, the administrative management implemented a radical (in a Danish context) reorganization of the public service for children with special needs, which affected both EDUADM and SOADM. In this case, the administrative management conceptualized the idea, decision and implementation of the innovation, which aimed at creating a new basis for improving collaboration between EDUADM and SOADM departments in order to decrease the number of children in special education and placement of children in care. On the part of the administrative management there was thus a strong incentive to change the public service area of children with special needs toward a reduction in the number of expensive cases combined with more control of a somehow 'uncontrolled' public service area.

The actual implementation of the new organization consisted in the formation of a new unit, called the Visitation Office (VIS), within the public service area for children with special needs. As the VIS was believed to ensure the realization of the innovation, it was perceived to be of pivotal importance by the administrative management. Before the reorganization in January 2005, the units under EDUADM and SOADM were responsible for normal schooling and preventive social and educational interventions as well as removals of children from families. After the reorganization the division of work changed. Now, the two administrations and their units were only committed to work with preventive intervention keeping the child in its natural environment, while the VIS was in charge of checking whether cases sent from the preventive system really were cases that required removal of the child, and in that case finding the appropriate solution.

At the outset, the idea behind the innovation was clear and simple: when a child with special needs was 'detected' by the involved members, the solution to the case should be based on coherent and effective collaborative actions by members of both the SOADM and the EDUADM. To give an example, if a school teacher noticed behavioural difficulties in a child, the teacher was required to contact and collaborate with, for instance, a social worker from the social service unit to try to find a solution among different types of preventive initiatives in order to avoid passing on the case to the VIS. The assumed perception underlying this innovation of the work process was that collaborative activities between organizational members from different units would prevent or reduce the number of cases passed on to the VIS. As a consequence, the public service area for children with special needs would be able to reduce the number of expensive solutions.

In case 1, 'The new office', the main goal for management was to improve the economy by reorganization aimed at improving collaboration between departments, units and individuals in the field. At that time, in a Danish context, the reorganization represented an innovative way of organizing the public service area for children with special needs. We define, by drawing on Schumpeter, the attempted reorganization as an example of innovation in which management attempted to put the productive resources of the two administrative offices to a use hitherto untried in practice and, to some extent, withdraw the two offices from the purposes they used to serve. What management did not take into account, however, was that what appeared to be innovation emerging from the practice-based learning of management was seen as a breach of important values to the other local workplace cultures linked to the practices of EDUADM and SOADM. The attempted innovation failed, creating instead what EDUADM and SOADM considered a chaotic situation (Brandi, 2010).

As the members of EDUADM and SOADM never recognized the attempted innovation, it was perceived as yet another obstacle to be dealt with in their everyday practices rather than being a means for a new shared focus on 'children with special needs'. This potential innovation was and remained an unsuccessful and unrealized attempt at innovation.

The employees in EDUADM saw children with special needs as individuals. These individual children had behavioural problems and learning disabilities and could thus be perceived as problems for the teachers, whose task it is to sustain and conduct everyday school activities. For the sake of the 'difficult' children as well as, and not least, the other children, EDUADM employees would work towards a fast placement rather than a prolonged prevention strategy. Their main motive was not reduction of costs related to children with special needs, but rather the maintenance of 'normal' classes at school.

The SOADM culture did not as such share the management's motive of reducing costs in relation to children with special needs. Their motive could more precisely be defined as 'helping families with special needs'; in so far a child was part of such a family it should be helped in its own environment, and therefore prevention was preferred over placement. A merger between the management culture and the SOADM culture thus appeared more straightforward. However, from the point of view of the SOADM culture, VIS was not a legitimate innovation, but an attack on their primary motive; to serve families with social problems and help children within the confines of that family life.

Though management maintained, on a discursive level, that the work of VIS was in the interest of children with special needs (as well as a reduction of costs and control), the employees in VIS were unable to form a common coherent motive. The psychologists in the office continued to work from their original motive of the EDUADM culture, and the social workers continued to work from their original SOADM culture motive. In the two remaining EDUADM and SOADM cultures, the only change was that the actual placement procedure, with its inherent conflicting motives, came under the responsibility of the VIS office. As the employees stuck to their original, local organizational culture, which originated from learning particular values, artefacts and emotions that came from practice-based learning, the attempted management innovation failed.

Case 2: 'new ways of lifting patients'

The second case also concerns potential innovation, which was never recognized by the organizational culture. It is the kind of potential innovation that is so hard to study in practice. In fact, this case grew out of private conversations between a researcher and a nurse, who was employed at a Danish workplace where her ideas were never recognized as innovative, because she never dared to tell the local management what she had done for fear of reprisals.

The nurse worked with old and troubled patients confined to their beds. She was deeply concerned with the patients' situation and able to put herself in their place. As part of their everyday tasks, the nurses had to hoist the bedridden patients out of their beds to change the linen every time something was spilled or the linen had to be changed for other reasons. The process was enormously complicated for all the nurses, and there always had to be more than one nurse to operate the hoisting apparatus. It was a complicated and time-consuming task, with braces and many straps being tied around the patient and to the apparatus.

The local nursing officer was very strict and demanded that everything must be done 'by the book' – just as she always stressed how she had to report any inaccurate behaviour to her superiors. One day a patient spilled some milk on the bedding, and, as there were no other nurses to help with the hoisting apparatus, the nurse resorted to an untraditional means of solving the situation instantly without too much bother for the patient. She grasped a big, black bin liner (made of very strong plastic) from the kitchen and carefully rolled it underneath the patient.

From this position it was easy to carefully push the patient to the foot of the bed. She removed the sheet from the top, replaced it with a new one, pulled the patient up on the new sheet, removed the dirty one from the foot end of the bed and tugged the new sheet in on the sides. The entire operation was completed without any colleagues present and took less than ten minutes. This episode has never been exploited as a potential innovation in nursing. On the contrary, the nurse would do anything to prevent the incident from being known for fear of dismissal by the nursing officer for breaking the rules. It only came to the researcher's attention through conversation with another employee and thus was not included as research data.

Methodologically, the case also shows how difficult it can be to study bottom-up processes of potential, emerging employee-driven innovation, especially if the potential innovation is never recognized or realized as an innovation.

Case 3: 'realizing the good idea'

The third case also concerns nurses who used their practice-based learning about patients' suffering to create new artefacts. In this case, the nurses' ideas were eventually recognized as innovative. Elise Soerensen was a nurse, who was born in 1904 and died in 1977. Throughout her career, she was recognized as a very devoted nurse who worked in the home care sector. In her work as well as in her private life she worked from a deep understanding of what it is like to be elderly and sick. To her, work was never just 'something-to-be-done' but a practice-based experience which built up values, artefacts and emotions of understanding and caring even when disease and symptoms of old age were hard to bear. These perspectives gained even deeper force when her sister Thora in 1954 was diagnosed with colon cancer and had a colostomy operation. In those days, keeping appropriate sanitation after such an operation was problematic, as patients often had to use dirty cloth bandages. Even more problematic was that the solution did not protect the patient from bad odour and leaks. As a consequence, most patients refrained from social contact after the colostomy operation.

Moved by her sister's difficult situation, Soerensen tried to develop a technical aid which could remedy the existing tools and came up with a radical innovation: a disposable plastic bag with an adhesive capacity to be placed directly on the body with an enclosed passage from body to bag protecting the patient from dirty linen, smell and leaks.

This is a classic situation of the creative innovative worker who comes up with a good creative idea with no organization to support its development into an organizational innovation. Instead, Elise designed and patented the product herself and went to the Danish producer of plastic products, Aage Louis-Hansen. This producer was more than sceptical when a nurse presented an idea so far from his own practice-based learning experiences. It was through his wife Johanne Louis-Hansen, who understood the situation of the patients, that Elise was finally able to persuade Aage to initiate the production of plastic colostomy bags. Initially 1,000 were produced, and they became an instant success. Elise distributed the colostomy bags to colleagues, who used them immediately, and Aage Louis-Hansen's company soon had difficulties keeping up with the demand. In 1957, a whole new company was founded building on Elise Soerensen's idea, and soon it too became a great export success. Today the company, Coloplast, has more than 7,000 employees worldwide (www.coloplast.com/about/pages/aboutcoloplast.aspx). This story is told on the website of the company and builds on letters and interviews by Elise during her lifetime. Elise herself was never to benefit greatly from this explosive development.

Coloplast later tried to create a workplace culture that encourages innovations that are not dictated by management but emerge from bottom-up processes and are linked to the everyday work practice of the employees. An example of a culture that takes account of the employees' practice-based experiences is that of the nurse Tine Richter Friis, whose experience was that nurses were unable to treat painful pus-filled wounds in bedridden patients. Many of the bedridden patients suffered enormously from wounds, but the nurses could not help because the treatment was too painful to the patients. The patients' suffering had not been registered anywhere except in the nurses' practice-based learning experiences. Together with R&D developers at Coloplast, Tine developed the idea of making a plaster with encapsulated painkillers. In this case, management took pride in realizing the employees' creative ideas (based on their everyday work life experiences) as employee-driven innovation.

A typology of EDI in workplace cultures

We have presented three different cases of innovation in the workplace. The three cases do not only differ in terms of the workplace culture-determined recognition of the employees' practice-based ideas. They also differ in methodological terms, and point to the difficulties of studying employee-driven innovation at the stage before recognition and implementation. How, then, is EDI enacted in the three cases? We shall attempt a typology of the different ways employees' innovation can be recognized as practice-based in an organizational culture.

Top-down management-driven EDI ignoring local practice-based cultures

The attempted innovation studied in case 1, 'The new office', led not only to the physical establishment of a new office but also to a number of tacitly obstructive actions and fierce negotiations concerning how employees should recognize the attempted innovation as added value and ascribe meaning to it. The VIS office as an innovation was enforced by the management and based on their notion that this new office would make former adversaries collaborate and thus reduce costs. This attempted innovation was bound to the managers' limited practice-based learning, and it could thus be argued that it failed because it lacked a shared culture. The management discursively constructed what was, from the perspective of the management culture, an innovative idea (i.e. to control and reduce the costs of children with special needs by promoting prevention and reducing placements), but never attempted to look into the actual practices of the two administrative offices, which had developed their own organizational cultures (with inherent emotions, discourses, etc.) supporting a different set of workplace practices and values. The management simply expected the employees to be innovative within the framework established and decided by the management. The everyday practice-based learning of the employees in EDUADM and SOADM was not taken into account and the values guiding their practical activities were thus ignored.

The above case illustrates that it may be difficult for management to demand that the employees truly become creative co-creators of innovation in the organization if the ideas presented by management do not recognize the values, artefacts, language and emotions the employees have already experienced and adopted through their practice-based learning. Thus, our case illustrates that innovation originating from top-down management-driven organizational processes requires more than simply a neat change program implemented by the top management. Innovation is highly dependent on the experiences and knowledge of the employees. It is, after all, these employees who are faced with children with special needs in the everyday activities of their work life. An innovation created and initiated in the management culture (claiming to have a 'god's eye view' over the whole organization) will be an answer to problems as they are defined in the management culture, not as the problems are experienced by the employees in their everyday life.

In the EDI literature we find many examples, from specifically appointed R&D offices to cooperative entrepreneurships, of top-down management-driven EDI in which managers either invite employees to contribute with ideas or try to impose innovation initiatives on their employees' practice-based work life. Such initiatives can be either successful or failed attempts by the management to identify sustainable ideas. The recognition of what is defined as innovation and what can be identified as a productive resource put to uses hitherto untried in practice is defined by the management and thus limited to the values of the management culture, which may be to a greater or a lesser extent in balance with the employees' everyday work life.

Unrecognized potential, bottom-up EDI created in local practice-based cultures

Following the distinction between creativity and innovation in the EDI literature, we contend that employees get many creative ideas that are linked to their everyday work life. Many of these ideas remain unrecognized because they are tied to the cultural values, artefacts and emotions of the employees' practice-based learning, which is neither known nor respected by management. Our second case 'New ways of lifting patients' further supports this statement. When the nurse thinks of using a plastic bag as a means to move her patients when cleaning or changing the dirty bedding, she knows she is not acting according to the rules in her department. Nevertheless, she has generated and put into use an idea which will help her move the patients without using a complicated and time-consuming machine that requires the presence of more than one nurse. Yet, according to the nurse, her idea can never be an institutionalized practice or routine since it is not a legitimate action, and in her department everything must be done 'by the book', meaning a strict protocol for acceptable and unacceptable behaviour is enforced by the nurse officer. As a consequence, the idea will never be recognized as an employee-driven innovation.

What is interesting here is that the attempt to make the individual innovation collectively known is not an issue she considers, since it collides with the values and norms espoused by management. The top-down organizational culture of the department thus determines what is legitimate and illegitimate behaviour, which is deeply inscribed in the practice of the employees. Again, management defines what is recognized as an innovation with no interest in how the employees' creative ideas, linked to local practices, can become innovation drivers of the organization as a whole. These 'could-be' employee-driven creative solutions to everyday problems, which hold the potential of being implemented as EDI, are difficult to study precisely because they pose an unacceptable risk of changing (and challenging) the defined top-down management-driven notion of the shared workplace culture.

Recognized bottom-up EDI, where the overall organizational culture acknowledges the creativity of local practice-based cultures as a basis for general innovation

The last case, 'Realizing the good idea', is a plain and successful example of an innovation process from idea generation to implementation and commercialization of a new product. Even though Elise Soerensen's idea is not accepted instantly, after some negotiation the idea leads to the foundation of the renowned healthcare production company, Coloplast. What is interesting about the third case is that the innovation is driven by the nurse Tine Richer Friis. Here, we observe an example of an organizational cultural recognition of an idea as a potential innovation stemming from the practice-based learning of Tine. In this case, the practice-based experiences, from local workplace cultures, of nurses who treat bedridden patients are recognized as valuable to the overall organizational culture. In the theoretical section we defined organizational culture as the shared self-evident and taken-for-granted patterns of meaning underlying the inclusion and exclusion of ideas and people in the organization. In Coloplast, the knowledge and learning from practice intrinsic to the idea of Tine Richer Friis are accepted and included in the organization as a potential innovation. The potential of the idea is not something that has to be debated or negotiated, since the Coloplast organization is built on and thus respects the employees' innovative potential. For example, the company homepage states that the contribution of Tine exemplifies the spirit of the founder Elise Soerensen. As illustrated by this example, emotions constitute an inherent part of practice-based learning, just as does the meaning associated with the use of artefacts. Today, Coloplast often describes itself as a company building on the 'spirit' of Elise Soerensen and has, according to its own descriptions, never ceased to be open to employees with innovative ideas. Here, the strong relation between an organizational culture that is open to values linked to local practice-based learning and the concept of EDI is emphasized.

We acknowledge the importance of management and management culture and practice as having a significant impact on analysing EDI. However, what the three cases demonstrate is that management is in many ways determined by the overall organizational culture that the managers are engaged in. Often the managers appear unaware of this. The managers alone do not dictate the culture of organizations. Organizational culture emerges from the interplay of managers and employees and the management's respect for the employees' potential innovation capability linked to local practice-based learning. Consequently, local values, artefacts and emotions are prerequisite for bottom-up EDI.

The concept of EDI – possibilities and limitations

Common to the three presented cases as regards EDI is that the employees' personal practice-based learning is what makes a real difference. This learning also involves certain cultural values, artefacts and emotions that may be supported or opposed by the values espoused by management. Management, we argue, tend to build their recognition of innovative potentials on values, artefacts and emotions rooted in their own practice-based learning rather than that of the employees. The practice-based learning of the management becomes a crucial element in being open or closed to recognition of employee-driven innovation, when resistance or recognition is rooted in different local cultural practices with their inherent motives, values and emotions. In order to understand and explain innovation processes and results, we argue that the concept of organizational culture is of great importance, not simply as an influencing factor behind innovation but also as the learning context within which EDI will be recognized and realized.

In many ways, the theory of practice-based learning can be seen as an argument that management plays an insignificant role in initiating the employees' creative, insightful ideas, but what is their role in turning creative actions initiated in local workplace cultures into innovation? Amabile and her crew present a useful distinction between 'creativity', as processes initiated by the individual employee, and 'innovation', as creative actions recognized in the organization as potential innovations, like one of the EDI discourses we presented in the first section of this article.

> The social environment can influence both the level and frequency of creative behaviour ... [and] [a]ll innovation begins with creative ideas ... We define innovation as the successful implementation of creative ideas within an organization. In this view, creativity by individuals and teams is a starting point for innovation; the first is necessary but not sufficient condition for the second. (Amabile *et al.*, 1996: 1155)

Employees get many good ideas when they go about their practice-based task in or outside the workplace. These ideas are often closer to the main objective of the company than the practice-based experiences of the management, because the employees' ideas and practice-based learning tend to be closely connected to the core objective of the company. These practice-based learning experiences can be very valuable if company managements open up to the recognition of employees' innovative capacities.

We have presented three cases of EDI and tried to analyse their main content and how they relate to innovation driven by employees. We have asked how we can understand the concept of EDI, and we argue that organizational culture is of essential importance for realizing and understanding innovation initiated by employees. We also find it necessary to move towards a broader understanding of culture than claims of 'innovation cultures' or 'management culture' being especially supportive of process innovation in particular adding value to the organization. We still need a better understanding of what is meant by culture in organizations and how this connects with EDI. With the proposed definition of culture as connected to practice-based learning in workplace activities, we argue that for bottom-up innovations to be realized they must be recognized as innovations by the overall organizational cultural context.

Our analysis of the EDI cases in this article demonstrates that innovation is highly dependent on the practice-based learning processes connected to culture. These learning processes form understandings of artefacts, emotions, discourses, values, resources and so on as core elements, which can facilitate and/or create inertia in the process of realizing EDI.

Our contribution has demonstrated that the concept of EDI can be approached from many different perspectives and must contain many different analytical levels. It is thus not enough only to look at the employee as the creative element and the manager as the innovation gatekeeper. One conclusion of this contribution is that, to deal coherently with the concept of EDI, one must not disregard the potential emergence of EDI from various structures (micro, meso and macro) and processes in workplaces and societies. Here, management simply constitutes one key factor among many other relevant factors, such as the culture at the workplace. We thus agree with Høyrup (2010) when he argues that in 'order to define employee-driven innovation, the concept should be linked to the fundamental criteria and questions of the broader term of innovation'. Hence, it makes no sense to talk about EDI in an isolated context – all processes in the workplace are embedded in different interests, internally as well as externally.

We have pointed to an unresolved discrepancy in the present application of the concept of EDI, as it is presented in the existing EDI literature. One can say that the EDI concept is still in the phase of being established as a scientific concept. However, we wish to emphasize that, from the beginning, it is important to understand EDI as more than either a bottom-up process where focus is solely on the employee or a top-down process initiated by management. In the empirical examples in the special issue on EDI in *Transfer* (Hoyrup et al., 2010) and the report by Rambøll, EDI is primarily understood as an external process initiated by governments, unions or management. In this literature, the perception seems to be that management or governments can cause or control the creation of an organizational culture that will be able to facilitate EDI, instead of bringing to the fore the employees and their potential innovations linked to their local practice-based workplace cultures.

Approaching EDI in this instrumental way indicates a perspective that allows external agents to inflict their practice-based learning (in the form of their specific values and artefacts) upon the employees, who must obey or act according to the given structure. Is that EDI? Our answer is: 'No.' We wish to draw attention to the necessity of an organizational culture that is open to the potential of innovation driven by the employees, who develop practice-based knowledge, values and experience from solving work-related issues every day in the local cultures. If EDI is to encompass a true bottom-up perspective, the management cannot dictate that innovation to the employees (case 1), nor can the organization of the management intimidate the workers into hiding local values, problems and innovative solutions (case 2). Our theoretical discussion of the concept of EDI outlines a model in which organizations must listen to the employees and create an organizational culture that builds on recognition of the practice-based potential for innovation in the employees.

Note

1. The Danish State has explicitly reserved a pool of funding for supporting employee-driven innovation in 2008–2011.

References

Ahmed, P.K. (1998). 'Culture and climate for innovation'. *European Journal of Innovation Management*, 1(1), 30–43.

Alvesson, M. and Berg, P.O. (1992). *Corporate Culture and Organizational Symbolism: An Overview*. Berlin: Walter De Gruyter Inc.

Amabile, T.M. (1988). 'A model of creativity and innovation in organizations'. In Staw, B.M. and Cummings, L.L. (Eds) *Research in Organizational Behavior* (Vol. 10, pp. 123–167). Greenwich: JAI Press.

Amabile, T.M., Conti, R., Coon, H., Lazenby, J. and Herron, M. (1996). 'Assessing the work environment for creativity'. *Academy of Management Journal*, 39(5), 1154–1184.

Åmo, B.W. (2006a). 'Employee innovation behaviour in health care: the influence from management and colleagues'. *International Nursing Review*, 53(1), 231–237.

Åmo, B.W. (2006b). 'What motivates knowledge workers to involve themselves in employee innovation behavior'. *International Journal of Knowledge Management Studies*, 1(1/2), 160–177.

Arundel, A. (2007). 'How Europe's economies learn: a comparison of work organization and innovation mode for the EU-15'. *Industrial and Corporate Change*, 1175–1210.

Axtell, C.M. (2000). 'Shopfloor innovation: facilitating the suggestion and implementation of ideas'. *Journal of Occupational and Organizational Psychology*, 73(3), 265.

Brandi, U. (2010). 'Bringing back inquiry: organizational learning the deweyan way'. In Jordan, S. and Mitterhofer, H. (Eds) *Beyond Knowledge Management: Sociomaterial and Sociocultural Perspectives within Management Research*. Innsbrück: Innsbrück University Press.

Brown, J.S., Collins, A. and Duguid, P. (1989). 'Situated cognition and the culture of learning'. *Educational Researcher* 18 (American Educational Research Association): 32–42.

Chiu, Y.C. and Liaw, Y.C. (2009). 'Organizational slack: is more or less better?' *Journal of Organizational Change Management*, 22(3): 321–342.

Deal, T.W. and Kennedy, A.A. (1982). *Corporate Cultures*. Reading, Massachusetts: Addison-Wesley.
Engeström, Y. (2001). 'Expansive learning at work: toward an activity theoretical reconceptualization'. *Journal of Education and Work*, 14(1), 133–156.
Fagerberg, J. and Verspagen, B. (2009). 'Innovation studies – the emerging structure of a new scientific field'. *Research Policy*, 38(2), 218–233.
Feldman, S.P. (1988). 'How organizational culture can affect innovation'. *Organizational Dynamics*, 17(1), 57–68.
Fenwick, T. (2003). 'Innovation: examining workplace learning in new enterprises'. *Journal of Workplace Learning*, 15(3), 123–132.
Floyd, S.W. (1999). 'Knowledge creation and social networks in corporate entrepreneurship: the renewal of organizational capability'. *Entrepreneurship Theory and Practice*, 23(3), 123–144.
Geertz, C. (1973). *The Interpretation of Culture*. New York: Basic Books.
Gherardi, S. (2000). 'Practice-based theorizing on learning and knowing in organizations'. *Organization*, 7(2), 211–223.
Godin, B. (2006). 'The linear model of innovation: the historical construction of an analytical framework'. *Science, Technology and Human Values*, 31(6), 639.
Gonzales, C.B. (2008). 'The cultures of international management'. *International Business and Economics Research Journal*, 7(7), 95–114.
Gudeman, S. (1992). 'Remodelling the house of economics – culture and innovation'. *American Ethnologist*, 19(1), 141–154.
Gupta, A.K., Tesluk, P.E. and Taylor, M.S. (2007). 'Innovation at and across multiple levels of analysis'. *Organization Science*, 18(6), 885–897.
Hasse, C. (2008). 'Cultural body learning – the social designation of institutional code-curricula'. In Schilhab, T., Juelskjær, M. and Moser, T. (Eds) *Body and Learning* (pp. 193–215). Emdrup: The Danish School of Education Press.
Hasse, C. and Trentemøller, S. (Eds) (2008). *Break the Pattern. A Critical Enquiry into Three Scientific Workplace Cultures: Hercules, Caretakers and Worker Bees*. Tartu: Tartu University Press.
Hasse, C. (2011). 'A review of psychological anthropology'. *Mind, Culture and Activity*, 9(4), 199–221.
Hasse, C. and Trentemøller, S. (2011). 'Cultural work place patterns in academia'. *Science Studies*, 24(1), 6–23.
Hatch, M.J. and Schultz, M.S. (2004). *Organizational Identity: a Reader*. Oxford: Oxford University Press.
Hayton, J.C. (2005). 'Promoting corporate entrepreneurship through human resource management practices: a review of empirical research'. *Human Resource Management Review*, 15(1), 21.
Herbig, P. and Dunphy, S. (1998). 'Culture and innovation'. *Cross Cultural Management*, 5(4), 13–21.
Høyrup, S. (2010). 'Introduction: employee-driven innovation and workplace learning: basic concepts, approaches and themes'. *Transfer*, 16(2), 143–155.
Hoyrup, S. (2010). *Transfer*, 16(2), 131–283.
Janssen, O. (2000). 'Job demands, perceptions of effort-reward fairness and innovative work behavior'. *Journal of Occupational and Organizational Psychology*, 73(3), 287.
Kiurunen, A.M. (2009). 'Culture effect on innovation level in European countries'. *International Journal of Business Innovation and Research*, 3(3), 311–324.
Lave, J. and Wenger, E. (1991). *Situated Learning: Legitimate Peripheral Participation*. Cambridge: Cambridge University Press.

Lemon, M. and Sahota, P.S. (2004). 'Organizational culture as a knowledge repository for increased innovative capacity'. *Technovation*, 24(6), 483–498.
Lin, L.H. (2009). 'Effects of national culture on process management and technological innovation'. *Total Quality Management and Business Excellence*, 20(12), 1287–1301.
McLaughlin, P., Bessant, J. and Smart, P. (2008). 'Developing an organisation culture to facilitate radical innovation'. *International Journal of Technology Management*, 44(3–4), 298–323.
Martin, J. (1992). *Cultures in Organisations – Three perspectives*. London: Oxford University Press.
Martin, J. (2002). *Organizational culture: mapping the terrain*. London: Sage Publications.
Martins, E.C. and Terblanche, F. (2003). 'Building organisational culture that stimulates creativity and innovation'. *European Journal of Innovation Management*, 6(1), 64–74.
Mavondo, F.T. and Farrell, M. (2003). 'Cultural orientation: its relationship with market orientation, innovation and organisational performance'. *Management Decision*, 41(3), 241–249.
Mavondo, F.T. and Rodrigo, E.M. (2001). 'The effect of relationship dimensions on interpersonal and interorganizational commitment in organizations conducting business between Australia and China'. *Journal of Business Research*, 52(2), 111–121.
Morcillo, P., Rodriguez-Anton, J.M. and Rubio, L. (2007). 'Corporate culture and innovation: In search of the perfect relationship'. *International Journal of Innovation and Learning*, 4(6), 547–570.
Nicolini, D., Gherardi, S. and Yanow, D. (Eds) (2003). *Knowing in Organizations. A Practice-Based Approach*. Armonk, New York: M.E. Sharpe.
Oldham, G.R. (1996). 'Employee creativity: personal and contextual factors at work'. *The Academy of Management Journal*, 39(3), 607.
Parker, M. (2000). *Organizational Culture and Identity*. London: Sage Publications.
Rambøll Management (2006). 'Medarbejderdreven Innovation på Private og Offentlige Arbejdspladser' ('Employee-driven innovation in private and public enterprises'), Rambøll Management, Report.
Reich, R. (1991). *The Work of Nations. Preparing Ourselves for 21st-Century Capitalism*. New York Vintage Books.
Schein, E.H. (1992). *Organizational Culture and Leadership*. San Francisco: Jossey-Bass Publishers.
Schumpeter, J.A. (1928). 'The instability of capitalism'. *The Economic Journal*, 38(151), 361–386.
Smircich, L. (1983). 'Concepts of culture and organizational analysis'. *Administrative Science Quarterly*, 28(3), 339–358.
Tan, B., Lee, C.K. and Chiu, J.Z. (2008). 'The impact of organisational culture and learning on innovation performance'. *International Journal of Innovation and Learning*, 5(4), 413–428.
Wolfe, R.A. (1994). 'Organizational innovation: review, critique and suggested research directions'. *Journal of Management Studies*, 31(3), 405–430.
Wynder, M. (2008). 'Employee participation in continuous improvement programs: the interaction effects of accounting information and control'. *Australian Journal of Management*, 33(2), 355.

8
Employee-Driven Innovation Amongst 'Routine' Employees in the UK: The Role of Organizational 'Strategies' and Individual 'Tactics'

Edmund Waite, Karen Evans and Natasha Kersh

Governments worldwide seek to upgrade the 'basic skills' of employees deemed to have low literacy and numeracy, in order to enable their greater productivity and participation in workplace practices. A longitudinal investigation of such interventions in the United Kingdom has examined the effects on employees and on organizations of engaging in basic skills programmes offered in and through the workplace. Through the 'tracking' of employees in selected organizational contexts, Evans and Waite (2010) have highlighted ways in which the interplay between formal and informal workplace learning can help to create the environments for employees in lower-grade jobs to use and expand their skills. This workplace learning is a precondition, a stimulus and an essential ingredient for participation in employee-driven innovation, as workers engage with others to vary, and eventually to change, work practices.

Generating an interplay between formal and informal learning can help to create the environments for employees in lower-grade jobs to use and expand their skills in and through the workplace. This workplace learning can support participation in employee-driven innovation, as workers engage with others to vary, and eventually to change, work practices. In keeping with this approach, we define 'workplace learning' broadly in terms of 'that learning which derives its purpose from the context of employment...learning in, through and for the workplace' (Evans *et al.*, 2006: 9) rather than merely in terms of learning that takes place through formal provision in the workplace. We also employ an expansive understanding of 'innovation' that includes not only major transformations and breakthroughs in knowledge but also 'those tiny incremental improvements that often fall off the radar screen but whose effect over time and in cumulative form can still be significant' (Bessant, 2003: 3).

Companies that aim to expand and enrich job content in jobs at all levels are likely to find employees working to expand their capacities accordingly. However, those that send employees on 'basic skills' courses, only to return them to a job and work environment that provides no opportunities for their use, are likely to see the benefits of their investment eroded over time. To be effective, workplace initiatives designed to support learning and innovation have to be based on realistic evaluations of the contexts and balances of advantage at three levels: the socio-political and organizational level (including the regulatory frameworks that govern the wage relationship); the immediate workplace environments; and the employees' dispositions to learn (Evans et al., 2006). When these conditions are met, the interplay between formal and informal learning can be powerful in creating environments and stimuli for employee involvement in innovation. In order to facilitate sustainability, government intervention is best directed to redressing the 'market failures' inherent in companies' investments in workforce development by supplementing companies' and unions' own efforts in these directions rather than importing external courses disconnected from core organizational concerns.

In this chapter, the sustainability of programmes designed to upgrade the competences and participation of workers in low-graded jobs is examined in greater detail, exploring the key factors that facilitate and inhibit workplace provision. We draw on the metaphor of a social ecology of learning to explore the interrelationships between individuals and groups at policy and organizational levels, and combine this with Michael de Certeau's theoretical work on quotidian social practices in order to cast light on the diverse ways in which various forms of 'basic skills' provision have been put to use by learners.[1] The paper argues that national government-led strategies to upgrade employees' literacy, numeracy and language skills have generated a complex 'ecology of learning' at policy level whereby a byzantine and shifting funding landscape, with its concomitant bureaucracy and strong emphasis on credentialism, has militated against long-term sustainable provision. Those organizations that have managed to sustain provision have generally succeeded in integrating workplace courses within a broader 'ecology of learning' whereby there is both support and formal recognition for such provision within the organization as a whole. The development of literacy, numeracy and ESOL (English for speakers of other languages) courses within these organizations approximates (rather than fully complying with) the 'Whole Organization' approaches advocated by key development agencies. Although these recommendations represent an optimum strategy for developing the capacity of organizations to deliver long-term provision, the 'third-order' priority of learning within the workplace means that it is in practice difficult to establish sustainability in most organizations. In the final section of the paper, we focus on two companies that have recently sought to develop innovative modes of delivery in order to circumvent these challenges.

The UK context

In 1999 a government inquiry headed by Lord Claus Moser highlighted a national 'skills crisis' facing the UK in the form of major literacy and numeracy skills deficiencies amongst adults (DFEE, 1999).[2] The report acted as a significant catalyst for the launch in 2001 of the national 'Skills for Life' strategy, which channelled more than £5 billion towards ring-fenced funding for free literacy, numeracy and ESOL provision. An important component of this strategy entailed major investment in the funding of literacy, numeracy and ESOL provision in the workplace in the form of discrete literacy, numeracy and ESOL courses in the workplace, literacy embedded in IT courses, and literacy embedded in vocational and job-specific training, as well as learn direct 'Skills for Life' courses undertaken in online learning centres in the workplace.

The publication of the Leitch Review of Skills in 2006 (an independent review by Lord Sandy Leitch commissioned by the British government in 2004) and the subsequent launch of the Train to Gain initiative (which is both a brokerage scheme to provide advice to businesses across England and an elaborate training scheme to fund full Level 2 and 'Skills for Life' provision for adults within the workplace) have further extended UK policy emphasis on the significance of 'Skills for Life' workplace provision. Despite a dearth of evidence relating to the impact of basic skills courses on productivity (Ananiadou et al., 2003), the government's rationale for investing in 'Skills for Life' provision throughout this period has been largely economic in nature: the development of literacy and numeracy skills amongst lower-level employees is deemed to be a vital means of enhancing the UK's global economic competitiveness (Wolf and Evans, 2011: 15).

In order to pursue the key factors that facilitate and inhibit sustainable provision, this chapter draws on longitudinal data from the 'Adult Basic Skills and Workplace learning' project (2003–2008) together with recent findings from research that is being undertaken under the auspices of the LLAKES research centre (Centre for Learning and Life Chances in Knowledge Economies and Societies). The 'Adult Basic Skills and Workplace Learning' project was based on structured interviews with 564 learners in 53 organizations from a variety of sectors (including transport, local government, food manufacturing, engineering and health) as well as structured interviews with the relevant managers and tutors at the selected sites. Further research into 'Skills for Life' workplace provision, based on this original data set together with case study research into organizations that have recently established innovative modes of provision, is being undertaken under the Strand 3 (project 2) of the LLAKES research centre, which is concerned with the social, economic and cultural factors that influence and impede individuals' attempts to control their lives, and their ability to respond to and manage opportunities.

Sustainable 'Skills for Life' workplace provision: theoretical perspectives

In examining workplace provision, including the factors that facilitate and inhibit sustainability, it is important to take account of three scales of activity (Evans et al., 2006). At the 'macro' level, wider social structures and social institutions can be fundamental in enabling or preventing effective learning from taking place. This includes the legal frameworks that govern employees' entitlements, industrial relations and the role of trades unions as well as the social structuring of business systems (Whitley, 2000: 88).

At the intermediate scale of activity, the nature of the learning environment in the organization can expand or restrict learning (see Fuller and Unwin, 2004). Establishing cultures that support expansive learning environments is problematic. For most employers, workers' learning is a lower-order decision and not a priority. As Hodkinson and Rainbird have noted, first-order decisions concern markets and competitive strategy. These in turn affect second-order strategies concerning work organization and job design. In this context, workplace learning is likely to be a third-order strategy (see also Keep and Mayhew, 1999).

The application of social ecological approaches ranges from macro-level policy analysis (Weaver-Hightower, 2008) to the adoption of the 'learning individual' as the unit of analysis in social psychological research (Bronfenbrenner, 1979) or, more recently, in the context of life-course research (e.g. Biesta and Tedder, 2007). In considering these scales of activity, it is important to avoid assumptions about the straightforward dissemination of educational policy and instead explore the contestation, selective appropriation and interpretation of educational initiatives at the policy, organizational and individual levels. As Ball (1998, cited in Weaver-Hightower, 2008: 153) states, 'most policies are ramshackle, compromise, hit and miss affairs, that are reworked, tinkered with, nuanced and inflected through complex processes of influence, text production, dissemination and, ultimately, re-creation in contexts of practice.'

De Certeau's (1984) conceptual distinction between 'strategies' and 'tactics' is useful in allowing an exploration of the intended and unintended consequences of 'Skills for Life' educational initiatives at the organizational and individual levels (Waite et al., 2011). In his analysis of the uses to which social representation and modes of social behaviour are put by individuals and groups, de Certeau links 'strategies' with institutions and structures of power, while 'tactics are utilized by individuals to create space for themselves in environments defined by strategies.' Strategies are only available to subjects of 'will and power' because of their access to a spatial or institutional location that allows them to objectify the rest of the social environment: 'A strategy assumes a place that can be circumscribed as a proper (*propre*) and thus serve as a basis for generating relations with an exterior distinct from it (competitors, adversaries, "clienteles," "targets," or "objects of research)" (De Certeau, 1984: 35). Although individuals lack a space of their own from which to apply strategies, they remain active agents through ongoing tactical practices which continuously re-signify and disrupt the schematic ordering of reality produced through the strategic practices of the powerful.

Through his analysis of a variety of everyday practices such as talking, reading, moving about, shopping and cooking, de Certeau illustrates his claim that everyday life works by a process of 'poaching on the territory of others', recombining the rules and products that already exist in culture in a process of 'bricolage' that is influenced, but never wholly determined, by those rules and products. The act of reading a book, for example, is described as a silent or hidden process of production (a poiēsis) 'which makes the text habitable, like a rented apartment. It transforms another person's property into a space borrowed for a moment by a transient' (De Certeau: XX1). In keeping with post modernist literary criticism, de Certeau suggests that 'every reading modifies its object...one literature differs from another less by its text than by the way in which it is read' (De Certeau, 1984: 169). De Certeau's theoretical work underlines the importance of taking account of the subterranean significance of individual engagement with social ecological patterns of behaviour:

> The Greeks called these 'ways of operating' *mētis*. But they go much further back, to the immemorial intelligence displayed in the tricks and imitations of plants and fishes. From the depths of the ocean to the streets of modern megalopolises, there is a continuity and permanence in these tactics. (De Certeau, 1984: X1X)

As will be seen in more detail in the next section, the use of literacy and numeracy provision in organizational 'strategies' is not always congruent with the broader 'Skills for Life' national strategy, which privileges the economic goal of developing literacy and numeracy skills in order to raise productivity. At the level of the individual, learners have engaged with 'Skills for Life' provision in order to pursue a wide array of goals that relate to their diverse and shifting lifestyles. In de Certeau's terms, they have tactically employed knowledge and opportunities afforded by 'Skills for Life' provision by using them with respect to ends and references that go beyond the highly instrumental economic agenda that underpins these interventions.

Analysis of data from the 'Adult Basic Skills and Workplace Learning' and LLAKES research projects: organizational 'strategies' and individual 'tactics'.

Despite the economic rationale underpinning UK 'Skills for Life' workplace provision, the majority of personnel managers interviewed as part of the 'Adult Basic Skills and Workplace learning' (ABSWL) project cited the importance of boosting the general development of employees as the primary motivation for delivering 'Skills for Life' courses rather than the need to address deficiencies in literacy and numeracy skills. Underpinning this goal lay a variety of motivations relating to the need to boost staff morale, foster a positive company ethos and enhance corporate solidarity as well as addressing unequal access to training opportunities amongst lower-level employees. In many cases, the courses were regarded as a useful means of compensating employees for the frequently routine and menial nature of their work. A manager of a bus company in the East Midlands, for example, outlined the demanding and tedious nature of the drivers' work and stated 'We can't change the conditions so we are trying to find other ways to make them feel better about themselves, their job and ultimately the company.' Many employees spoke English as a second language (ESOL learners represented a sizeable 35 per cent of the full sample, whereas only 3 per cent of employees in the current UK workforce do not speak English as their first language) and there was interest in improved communication, though it was rarely seen as central to job performance.[3]

While companies and public sector organizations have utilized 'Skills for Life' provision to pursue a variety of 'strategic' objectives – relating largely to the need to develop the psychological contract between employer and employee – learners have 'tactically' insinuated an even more diverse array of goals and understandings into the experience of undertaking a literacy, numeracy and ESOL course in the workplace. Quantitative data from Time 1 and Time 2 of the ABSWL project revealed that employees were motivated to engage in workplace 'Skills for Life' provision by a far wider range of factors than merely the wish to improve performance at work. During the course of in-depth interviews, learners divulged in more detail a whole range of factors for engagement in such courses: from 'curiosity' to wanting to make up for missed earlier educational opportunities; from wanting specific help with job-relevant skills to wider career aims; from a desire to help children with school work to wanting self-improvement and personal development (Evans et al., 2009).

In-depth interviews also cast light on the wide range of individual and social strategies for coping with existing literacy and numeracy skills (e.g. reliance on colleagues and supervisors for support with form-filling) as well as the significance of 'informal learning' in developing these skills in a variety of workplace contexts.[4]

In this respect it is important to take account of the wide variation of literacy practices in differing organizational contexts; whereas some employees (e.g. care workers in residential care homes) remarked upon an increase in report-writing in response to auditing demands and more onerous health and safety regulations, the majority of employees were engaged in occupations that entailed the persistence of routine work in which there was negligible use of literacy and numeracy practices. Such findings are compatible with a growing corpus of research that has underlined the persistence of relatively routine or manual employment in large swathes of the UK economy (e.g. Felsted *et al.*, 2007; Keep, 2000; Keep and Mayhew, 1999; Lloyd *et al.*, 2008).

In de Certeau's terms, learners have effectively 'made of' the knowledge and skills afforded by workplace 'Skills for Life' provision – in a 'hidden' or 'secondary' process of production – by using them with respect to references and ends that relate to their diverse and frequently shifting lifestyles. Organizations' promotion of 'Skills For Life' courses for largely generic rather than job-specific considerations has provided a broad domain for the pursuit of individual 'tactics' which variously intersect with and diverge from company strategic objectives according to the complex interrelationship between learner-specific considerations and organizational imperatives.[5]

Case study analysis of organizations that have developed long-term 'Skills for Life' provision

Out of the 53 sites recruited for the 'Adult Basic Skills and Workplace Learning' Project, 11 of these are running 'Skills For Life' provision in 2010: four public sector organizations (two local authorities, one hospital, one public sector transport company), six private sector companies and one charity.

Only seven sites may be described as having reasonably durable provision during the project timescale, in so far as provision was running at the time of ABSWL Time 1, Time 2 and follow-up LLAKES research interviews. Four sites can be described as having 'intermittent' provision, in so far as courses were not running at the time of Time 2 interviews but had been revived at the time of LLAKES follow-up interviews in 2009–2010.

As argued elsewhere (Waite *et al.*, 2011), complex and rapidly shifting funding arrangements (which have placed a strong emphasis on target-bearing qualifications) in combination with a range of challenges related to adapting provision to the workplace environment have posed major barriers to longer-term provision. The negative impact of the economic downturn has further compounded the constraints on sustainability.

The organizations that succeeded in developing long-term provision throughout the duration of this timescale are summarized below.

Southern Transport Systems (STS) (a large transport provider with approximately 17,000 employees). The company has been running English, maths, IT and dyslexia courses at learning centres since the publication of the Moser report in 1999.

Thorpton Local Authority in London Union Learning Representatives (ULRs) have played an important role in implementing a series of 'Skills for Life' courses at Thorpton local authority's learning centre in partnership with the learning centre manager and local colleges.

Lindall plc (a food manufacturing company in Cornwall with 700 employees) has been delivering literacy courses for its shop floor workers as well as an ESOL course for Kurdish employees since 2003.

Brandon Care Home (a purpose-built village community for people with learning disabilities, employing 354 staff) has delivered a series of ESOL and literacy courses since 2004 for approximately 150 learners, consisting largely of care workers.

Melford Hospital has delivered literacy and numeracy courses for hospital employees during working hours and has recently combined this with online provision.

Finross City Council has delivered literacy, numeracy and ESOL courses for council employees since 1999 on the basis of funding from the Scottish Executive.

Baden plc (an 'asset management' company with 54,000 employees) combines 'Skills for Life' provision with job-specific training (in cleaning and security).

'Skills for Life' development agencies promote the importance of establishing 'whole organization' approaches to the development of 'Skills for Life' provision in the workplace and other sectors 'where consideration of literacy, language and numeracy (Skills for Life) provision is central to the whole organization at all levels, ranging from strategic leadership and management to the delivery of all services, including those involving training and development' (QIA, 2008: 10). The Skills for Life Improvement programme, run by the Quality Improvement Agency (QIA), recommends that large employers develop a variety of measures to achieve these aims, including the establishment of a steering group with representatives at all levels (including union support), the support of senior management commitment, the incorporation of Skills for Life priorities into policies, procedures and plans, and the undertaking of a training needs analysis and staff surveys (QIA, 2008).

Six out of the seven highlighted sites managed to implement provision that approximated (without fully complying with) these 'whole organization' approaches in so far as they benefited from the support of senior management, who were frequently enlisted in order to advocate the benefits of literacy training to the line managers of learners. The consolidation of SFL provision has also depended on the active involvement of unions (in the form of union learning representatives) in the case of STS, Thorpton, Melford and Finross City Council. Lindall plc represents an exceptional case in so far as provision within this company depended on the work of one personnel manager. Provision within this company would therefore appear to be most vulnerable, since it is entirely dependent on the hard work and determination of an individual 'key player'. In all these sites, 'Skills for Life' provision has been effectively integrated into an organizational 'ecology of learning' in so far as provision has effectively responded to a clearly identified need in the workplace as well as catering for the diverse interests and motivations of learners outside the workplace. In STS systems, courses have responded to increased report-writing in certain sectors of the organization as well as allowing learners to pursue generic interests (including courses that have allowed learners to undertake GCSEs in English). The two local authorities (Finross and Thorpton) have effectively used 'Skills for Life' provision to respond to the need for increased report-writing by employees and have established learning centres that provide accessible and non-intimidating spaces for learning. The courses at Lindall have also facilitated the capacity of employees to respond to the increasing 'textualization' (Scheeres, 2004) of the workplace (e.g. in the form of health and safety regulations) as well as fostering the personal development of employees. Training managers at Brandon Care Home have been adept at drawing on learning online in order to facilitate a more flexible 'ecology of learning' within this organization.

In embracing wide-ranging motivations for learning (rather than focusing exclusively on addressing literacy and numeracy skills deficiencies), these organizations have provided a broad institutional space for learners to tactically 'make of' their learning in order to pursue an array of personal goals. It is noteworthy that all these organizations are large employers with sufficient resources to uphold the wide-ranging benefits of learning and establish a robust learning infrastructure. This is consistent with research undertaken on US literacy provision in the workplace, which found that larger firms (those with more than 500 employees) were more likely to establish longer-term courses (Nelson, 2004).

As we have argued elsewhere, 'Skills for Life' provision can foster a positive interplay between formal and informal learning (Evans and Waite, 2010; Taylor et al., 2007). In Thorpton council, for example, the tutor and manager remarked upon an increased confidence on the part of employees which has led to the development of further formal learning opportunities (through willingness to embark on further learning) as well as informal learning opportunities (through the taking on of higher-level roles that also entail 'hands-on' learning). The development of sustainable 'Skills for Life' provision with its attendant synergies between formal and informal learning can in turn facilitate employee-driven innovation in the sense of the 'tiny incremental improvements' that Bessant argues are not always officially recognized. The increased capacity of station assistants to deal with the textualization of the workplace in STS systems, the greater capacity of caretakers to document cases of vandalism, the increased capacity of employees at Brandon Care Home and Melford Hospital to write care plans for the elderly as well as handover notes, facilitates the development of an archive of not only observations but also employee initiative in addressing workplace challenges which can be drawn upon to inform higher-level management planning and innovation. Equally, the greater facility of lower-level employees to use email and ICT more generally facilitates channels for suggesting changes to working practices and therefore allows enhancement of innovation at all levels of the organization.

'Skills for Life' workplace provision has also facilitated the processes of innovation that are inherent to the 'tactical' practices of everyday work. In addition to enlivening the workplace by allowing learners to 'make of' the courses in order to pursue a wide range of personal and work-related motivations, such provision has also equipped learners with greater resources for making their working lives more meaningful through enhanced literacy and ICT skills. The manifold outcomes of this provision include the enhancement of independent learning on computers in company learning centres (Evans and Waite, 2008: 9) as well as greater utilization of literacy for personal interest and development in the workplace (e.g. during breaks from work) as well as at home.

Forging new modes of delivering 'Skills for Life' workplace provision

In this section we focus on two organizations that have recently developed new modes of provision that eschew the traditional model of delivering workplace 'Skills for Life' through group sessions organized according to a fixed timetable. These organizations have been the subject of recent research as part of LLAKES Strand 3 project 2 into new modes of delivery that have sought to circumvent barriers to sustainable workplace provision. In the case of Leva, highly flexible forms of one-to-one delivery have been formulated in order to respond to the demanding shift patterns in a traditional factory setting. In the case of Fenton, a hotel and restaurant group, the use of new technologies as well as a network of tutors (entitled 'learning support managers') who work on a very flexible schedule have succeeded in tailoring provision to more than 1,000 learners in a large number of sites throughout the UK.

Leva car manufacturing company

The Leva factory in London makes parts for cars and employs 4,000 staff. Since 2007 a training provider has been delivering innovative 'Skills for Life' provision which entails specialized one-to-one individual tuition for factory staff next to the production lines. This model of delivery involves eight tutors teaching eight one-to-one 30-minute sessions a day, engaging a total of between 60 and 90 learners during the course of three different shifts (morning, afternoon and night).

The company has succeeded in training 1,000 learners (a quarter of the workforce) over three years (from 2007 to 2010). 'Skills for Life' provision at Leva, which is based on a combination of European Social Fund, London Development Agency and Train to Gain funding, benefits from broader company support; representatives from the training provider attend a 'Skills for Life' training committee which also includes union representatives as well as senior management.

The individual sessions take place in 'rest areas' (i.e. seating areas where the operatives normally take their breaks) near the production lines. The group leaders provide cover for the operatives whilst they undertake the 'Skills for Life' sessions. There is also a company learning centre with two computers that are available for self-study by learners. The tutors rely on laptops for the use of initial assessments and tests, as well as a combination of paper-based and ICT resources for teaching. The training manager states that, since the workforce is made up of a large number of older employees, many of whom do not speak English as a first language, sole reliance on ICT provision would not be feasible: 'But the fact is, you can't just leave a learner – and also a 'Skills for Life' learner – some of them just haven't got the motivation to be... or the skill even to be able to sit at the computer. I mean we're finding people here that are frightened to even switch the thing on, or don't even know where to switch it on.'

The 'Skills for Life' training responds to literacy requirements in the workplace, including risk assessments, health and safety regulations and the increasing delegation of QPS (quality production system) responsibilities to production operatives. The courses also cater for the wider personal development needs of the employees. In addition, the courses have responded to a growing company emphasis on lower-level employee involvement in decision-making processes and therefore facilitate channels for employee-driven innovation. As the training manager reveals: 'Whereas at one time group leaders were the ones that actually put in any new things on the line – now it's expected that the operators themselves, you know, look at their workstation, look how it's working and think "can they make amendments". That can't be done unless it's through the group leader but they're expected to think about it themselves, whereas at one time that was never expected.'

Fenton

Fenton is a large hotel and restaurant group, employing over 33,000 people in 650 sites across the UK. Since signing up to the 'Skills Pledge' in 2007, Fenton has become committed to developing a range of initiatives facilitating efficient 'Skills for Life' delivery across all its sites. In collaboration with a private training provider (specializing in technology-enabled learning solutions and services), the company has succeeded in delivering 'Skills for Life' provision to more than 1,000 learners through a discrete SFL course as well as through an embedded apprenticeship programme, which includes training for qualifications in four job roles: housekeeping, food and drink service for restaurant and bar, kitchen, and reception.

Both courses take the form of virtual learning delivered through an 'el-box' (a touch screen tablet PC), which is available at all the sites. The key benefits of using the el-box in the workplace include convenience and accessibility, as well as providing opportunities to deliver in-house, on-the-job training courses, thus eliminating the inconvenience and cost of sending learners to off-site classroom-based courses.

The learners are fully supported by a team of professional Learning Support Managers (LSMs) who provide the necessary on-site additional guidance, support, feedback and assessment for learners across all Fenton sites through email, telephone and face-to-face support.

The content can also be accessed from a website, thus providing the learners with an even greater degree of flexibility, enabling them to carry on with their 'Skills for Life' courses from their home or any other settings outside their workplaces. The interviews with the learning support managers have indicated that this kind of flexible arrangement is considered to be beneficial for the learners, as it responds to their individual needs and circumstances:

> obviously they have the opportunity to do it within work but sometimes people feel a little bit more comfortable doing this in their own environment. And it's about them putting a little bit in for their learning. [Extract from manager's interview]

This mode of delivery underlines the benefits of flexible training that allows employees to shape their own learning agendas around their needs and circumstances at work and home.

Conclusion

The complex and shifting funding landscape, with its strong emphasis on target-bearing qualifications. combined with a range of challenges related to the 'third-order' significance of learning in the workplace, has posed major impediments to sustainable provision at the organizational level. Those organizations that have succeeded in sustaining provision over the longer term have generally managed to harness broader support for this type of provision within the organization as a whole in a manner that approximates the 'whole organization' approaches advocated by 'Skills for Life' development agencies. These organizations have upheld the generic benefits of learning (for a variety of company-specific and altruistic reasons) whilst responding to a specific need for the development of literacy and numeracy in the workplace, thereby generating the potential for the positive interplay between formal and informal learning which in turn can potentially widen the channels for employee-driven innovation.

The narrowly focused economic agenda underpinning the 'Skills for Life' national strategy is at variance with the widely ranging motivations underpinning organizational and individual engagement with literacy provision. Government declarations of a 'skills crisis' based on assumptions about the existence of large-scale deficiencies in literacy and numeracy skills amongst lower-level employees have taken insufficient recognition of the variation of literacy practices amongst lower-level employees in differing organizational contexts as well as the complex constitution of employee skills and competencies, which frequently rely on 'informal' methods of learning in differing workplace settings.

We have suggested that Michel de Certeau's work on quotidian social practices – and in particular his conceptual distinction between 'strategies' and 'tactics' – provides a useful theoretical framework for understanding the processes of adaptation and accommodation entailed in the implementation of this type of provision. Rather than explicitly subverting or rejecting dimensions of 'Skills for Life' provision, learners have 'made of' the opportunities afforded by the workplace provision by using them with regard to ends and references that extend beyond the workplace and relate to their diverse and shifting lifestyles in a manner that frequently compensates for previously negative educational experiences. In their bid to boost staff morale, foster a positive company ethos and enhance corporate solidarity, the majority of organizations have developed a broad 'strategic' terrain in which individuals have been able to deploy these 'tactical' engagements more extensively. Most of the institutions that have managed to sustain long-term provision have accorded official recognition to the value of learning for its own sake (for a variety of company-specific as well as altruistic considerations) and have provided an institutional space for learners to pursue diverse interests and motivations in addition to addressing skills requirements in the workplace.

By drawing on the concept of 'poiēsis', de Certeau highlights the acts of production that take place in everyday social life but which are not always formally recognized, remaining 'invisible' or 'secondary'. In challenging the associations of action and passivity that are implicit in the producer/consumer binary, de Certeau's work also facilitates, in the context of workplace learning, the scrutinizing of a range of binaries – for example, teachers/learners or knowledge workers/routine workers – that can all too frequently depict the latter half of these distinctions as the passive recipients of those who generate and disseminate new forms of knowledge or instil new working practices. Even those 'routine' workers who appear not to engage in the formation of new types of knowledge and skills may engage in innovative work practices. This occurs not only because lower-level employees can make suggestions for small-scale improvements that may inform higher-level planning and innovation but also because they deploy various 'tactical' practices that are designed to enhance the quality of their working lives (e.g. through developing social ties with colleagues). As seen above, Skills for Life provision has important implications in terms of facilitating the channels of employee-driven innovation as well as equipping learners with resources with which they can 'tactically' make their working lives more meaningful.

We have highlighted two case studies of organizations that have forged new modes of delivery that are based on a large degree of institutional flexibility (on the part of organizations and providers), which has in turn facilitated the adaptive potential of the courses for the personal and work requirements of employees. Leva has relied on a model of one-to-one delivery that seeks to overcome the difficulty of adjusting provision to shift patterns. This mode of delivery has facilitated employees' capacity to fulfil literacy requirements relating to quality monitoring and health and safety. The provision has also responded to company expectations that operatives should take the initiative in relation to innovating more effective working practices on the production line. Fenton has drawn on new technological devices and an elaborate network of tutors in order to provide extremely flexible patterns of delivery that cater for employees' work-specific and personal requirements. Such modes of delivery present innovative attempts to circumvent the range of challenges related to adapting provision to the workplace that have, in combination with funding constraints, posed major barriers to sustainable workplace 'Skills for Life' provision.

Funding: This work was supported by the Economic and Social Research Council (ESRC Award RES-139–25–0120 and Award RES-594-28-0001).

Notes

1. For an analysis of 'social ecological' approaches to learning, see Evans *et al.* (2010).
2. See OECD (1997) for a description of the IALS data on which the Moser committee based its recommendations. More recently, the Skills for Life needs survey of 2002/03 produced a lower indication of those adults who struggle with literacy, with an estimate of 5.8 million people below Level 1 (Williams, 2003).
3. Employers underwrote participation in paid working time. All sites incurred organizational costs, not least in negotiations with line managers over shifts; many provided equipment and furnished teaching space.
4. See Evans and Waite (2008; 2010) for an analysis of the interviewing of 'formal' and 'informal' learning opportunities related to workplace Skills for Life provision.
5. This research also found that courses that continued over the long term (after initial funding had ceased) shared the common features of: 'an internal champion who had decision-making power or knew how to influence those who did, a well-identified internal issue or problem, and evidence that the program had helped to address that issue.'

References

Ananiadou, K., Jenkins, A. and Wolf, A. (2003). *The Benefits to Employers of Raising Workforce Basic Skills Levels: A Review of the Literature*. London: National Research and Development Centre for Adult Literacy and Numeracy.

Ball, S.J. (1998). 'Big policies/small world: an introduction to international perspectives in education policy'. *Comparative Education*, 34(2), 119–130.

Bessant, J. (2003). *High-Involvement Innovation: Building and Sustaining Competitive Advantage through Continuous Change*. West Sussex: John Wiley and Sons.

Biesta, G. and Tedder, M. (2007). 'Agency and learning in the life course: towards an ecological perspective'. *Studies in the Education of Adults*, 39(2), 132–149.

Bronfenbrenner, U. (1979). *The Ecology of Human Development: Experiments by Nature and Design*. Cambridge: Harvard University Press.

De Certeau, M. (1984). *The Practice of Everyday Life*. Berkeley: University of California Press.

Department for Education and Employment. (1999). *A Fresh Start: Improving Literacy and Numeracy. The report of the Working Group chaired by Sir Claus Moser*. London: DfEE.

Department for Education and Employment. (2001). *'Skills for Life' – The National Strategy for Improving Adult Literacy and Numeracy Skills*. London: DfEE.

Evans, K. and Waite, E. (2008). 'Adult workers' engagement in formal and informal learning: insights into workplace basic skills from four uk organisations'. Ottawa, Ontario, Canada: Partnerships in Learning. Web publication: http://www.nald.ca/library/research/interplay/insights/insights.pdf [accessed 4 March 2010]

Evans, K. and Waite, E. (2010). 'Stimulating the innovation potential of "routine" workers through workplace learning'. *Transfer: European Review of Labour and Research*, 16(2), 243–258.

Evans, K., Hodkinson, P. and Unwin, L. (2002). *Working to Learn: Transforming Learning in the Workplace*. London: Kogan Page.

Evans, K., Hodkinson, P., Rainbird, H. and Unwin, L. (2006). *Improving Workplace Learning*. London and New York: Routledge.

Evans, K., Waite, E. and Admasachew, A. (2009). 'Enhancing "skills for life"? adult basic skills and workplace learning'. In Bynner, J. and Reder, S. (Eds) *Tracking Adult Literacy and Numeracy: Lessons from Longitudinal Research*. London and New York: Routledge.

Evans, K., Waite, E. and Kersh, N. (2010). 'Towards a social ecology of adult learning in and through the workplace'. In Malloch, M., Cairns, L., Evans, K. and O'Connor, B. (Eds) *The Sage Handbook of Workplace Learning*. London: Sage Publications.

Felstead, A., Gallie, D., Green, F. and Zhou, Y. (2007). *Skills at Work 1986–2006*. University of Oxford: SKOPE.

Fuller, A. and Unwin, L. (2004). 'Expansive learning environments: integrating organisational and personal developments'. In Rainbird, H., Fuller, A. and Munro, A. (Eds) *Workplace Learning in Context*. London and New York: Routledge.

Hodkinson, P. and Rainbird, H. (2006). 'Conclusions: an integrated approach' (chapter 8). In Evans, K., Hodkinson, P., Rainbird, H. and Unwin, L. *Improving Workplace Learning*. Routledge: Abingdon.

Keep, E. (2000). *Learning Organisations, Lifelong Learning and the Mystery of the Vanishing Employers*. Working Papers of the Global Colloquium on Supporting

Lifelong Learning [online]. Milton Keynes, UK: Open University. Available from http://www.open.ac.uk/lifelong-learning [accessed 2 February 2010]
Keep, E. and Mayhew, K. (1999). 'The assessment?: knowledge, skills and competitiveness'. *Oxford Review of Economic Policy*, 15(1), 1–15.
Lloyd, C., Mason, G. and Mayhew, K. (2008). *Low-Wage Work in the United Kingdom*. New York: Russel Sage Foundation.
Nelson, C. (2004). 'After the grant is over: do workplaces continue to fund programs that were initiated with public funds'. *Focus on Basics: Connecting Research and Practice*, 7(2), 1–6.
OECD (1997). *Literacy Skills for the Knowledge Society*. Paris: OECD.
Quality Improvement Agency (2008). *Whole Organisation Approaches to Delivering 'Skills for Life': Making the Case*. Available from http://www.excellencegateway.org.uk/page.aspx?o=159798 [accessed 5 October 2010]
Scheeres, H. (2004). 'The textualised workplace'. *Reflect*, 1, 22.
Taylor, M., Evans, K. and Abasi, A. (2007). 'Understanding teaching and learning in adult literacy practices in Canada and the United Kingdom'. *Literacy and Numeracy Studies: An International Journal in the Education and Training of Adults*, 15(2), 57–72.
Waite, E., Evans, K. and Kersh, N. (2011). *Is Workplace 'Skills for Life' Provision Sustainable in the UK?* Published by the Centre for Learning and Life Chances in Knowledge Economies and Societies at: http://www.llakes.org.
Weaver-Hightower, M.B. (2008). 'An ecology metaphor for educational policy analysis: a call to complexity'. *Educational Researcher*, 37(3), 153–167.
Whitley, R. (2000). *Divergent Capitalisms: The Social Structuring and Change of Business Systems*. Oxford: Oxford University Press.
Williams, J. (2003). *The 'Skills for Life' Survey: A National Needs and Impact Survey of Literacy, Numeracy and ICT Skills*. London: The Stationery Office.
Wolf, A. and Evans, K. (2011). *Improving Literacy at Work*. Abingdon: Routledge.

Part III
Employee-Driven Innovation Unfolded in Global Networks and Complex Systems

9
Moving Organizations towards Employee-Driven Innovation (EDI) in Work Practices and on a Global Scale: Possibilities and Challenges

Maja Lotz and Peer Hull Kristensen

Exploring 'organizational moves' towards EDI within the field of Danish multinationals

Today, multinational corporations (MNCs) increasingly innovate by tying into global networks of customers, suppliers, public R&D institutions and other external partners (Bartlett and Ghoshal, 1989; Hedlund, 1986; Sabel *et al.*, 2009), experimenting with new ways of combining and leveraging distinctive knowledge and practices from around the world. Innovating in global networks, MNCs can no longer depend on cues from a centralized R&D lab or top management. Nor can they rely on former 'transfer' or 'projection' strategies to introduce innovations and advantages abroad through one-directional transfer of resources, information and knowledge from home to overseas environments (Chesbrough, 2003; Doz *et al.*, 2003). To improve their innovative performance, instead, they are required to combine knowledge and learn from multiple sources (embedded in diverse organizational and institutional contexts) as well as to decentralize responsibilities and innovative search practices to various levels and sites that can respond quickly to new situations. This becomes so much more important as the firm constantly has to define for itself a new role in relation to other firms, with constant changes in the composition of global value chains, role changes that make it necessary to reform the organization, focus on new processes and bring new products to the market (Herrigel, 2010). Consequently, not only R&D units but organizational members at all levels (i.e. employees, managers, suppliers, customers and other partners) of MNCs and their surrounding institutions need to engage in mutual knowledge-sharing and an ongoing distributed search for innovations in and across various 'collaborative communities' (Dorf and Sabel, 1998; Heckscher and

Adler, 2006; Sabel, 2007; Stark and Girard, 2002; Wenger, 1998). In response, they have developed new ways to collaborate and share knowledge and decentralized innovations in and across different groupings, work teams and units. Such practice-based forms of innovation reflect the ability of organizational members to actively take part in, and continuously innovate through, processes of joint learning during daily work activities.

Prior research has shown how these non-R&D-based innovative forms are crucial for firms' innovative capabilities, as they magnify the ability to accumulate fresh knowledge and co-create in innovative ways among organizational members. Yet, to date, we have a limited understanding of how practice-based forms of innovation are in fact organized, distributed and managed at the micro-level of everyday work-organizing practices – that is, among organizational members interacting in and across various collaborative team settings. In this chapter we provide insights into these poorly understood aspects of innovation by studying how Danish multinationals, through experimentation with new organizational forms, have built up highly collaborative work-organizing practices facilitating processes of 'everyday' innovation throughout the organization. We focus on one particular form of practice-based innovation, namely employee-driven innovation. Conceptually, we understand employee-driven innovation as habitual routine-breaking (mental and physical) search activities among employees aimed at exploring and better exploiting organizational resources for further innovative recombination (March, 1991; Stark, 2009; Weick, 1979) – for example, when operators develop new products and processes in close collaboration with construction, search for better technical solutions with machine suppliers or find novel solutions for customers by sharing and recombining knowledge and practices in renewing ways. Within this view, innovation is conceived as a process – a flow of routine-breaking activities/ interactions – rather than as a static or fixed product-orientated phenomenon. This conceptualization implies two claims: first, that action is social, reflexive and ultimately creative. When our interactive habits, for instance, are disrupted through the course of social life, we seek to repair our relations by reconceiving them (Herrigel, 2010; Joas, 1996; Mead, 1934). Such imaginative interaction causes unforeseen possibilities for action to emerge and thus the potential for new creative activities and innovative learning. Second, our conception therefore also assumes that learning – understood as reflective creative experience (Dewey, 1938; Pierce, 1877; 1992) – is a core ingredient of practice-based innovative activities. Learning is, in other words, an integral part and a crucial enabler of all innovative (inter)actions. Based on this conceptualization we look into how firms seek to organize for, distribute and govern employee-driven innovation in and across local and global teams.

Empirically, we draw on findings from a larger study of seven Danish manufacturing companies which (1) are all part of a multinational and

(2) have built up highly collaborative, constantly redefining and laterally accountable team-based work-organizing practices (Kristensen and Lotz, 2011; Lotz, 2009). The chapter is based on our ethnographic studies within two of the seven companies conducted from 2006 to 2008. Informed by this empirical landscape, we explore four distinct 'organizational moves' towards employee-driven innovation. By 'organizational moves' we refer to a shift of organizational structures and work practices that in particular seem to facilitate dynamics of employee-driven innovation. The four 'moves' are selected for investigation because they are particularly rich in information on the development of work-organizing practices that enable employees to take on the role of 'drivers' of innovation.

The chapter proceeds as follows. First, we briefly describe how innovation studies have taken a new course, directed towards work organization at the micro-level and combined with a global view of networks at the macro-level. Then we investigate in depth two companies that illustrate this trajectory and study their transformation as four 'organizational moves'. We conclude this central part of our analysis by relating the moves to the institutional context and legacy in which they took place, and ask whether this could also happen under different institutional conditions. Finally, we suggest that a clarification of this question could be achieved by studying how firms with elaborated forms of EDI try to configure foreign subsidiaries, based on involving employees much more actively under very different institutional conditions than in their home economies.

Moving towards new organizational understandings and practices of innovation

Prior conceptualizations of innovation have focused on macro phenomena such as the creation of economic value (Schumpeter, 1934), development of new products (Cooper, 1990; Pianta, 2005; Trott, 1998), technology (Carlsson and Stankiewicz, 1991), user-driven innovation (Jeppesen, 2005; TemaNord, 2006; Von Hippel, 1988), a distributed innovation network (Bartlett and Ghoshal, 1989; Hedlund, 1986), national innovation systems (Christensen, 2003; Edquist, 2005; Lundvall, 1992; Whitley, 2006) or open innovation (Chesbrough, 2003). Less attention has been given to non-R&D-based innovative forms, for instance practice-based forms such as employee-driven innovation or day-to-day innovative learning, which take place not only in and between R&D units but throughout organizations (Ellström, 2006; Fenwick, 2003; Jensen *et al.*, 2007; LO, 2007), or to how such innovative practices may be distributed and governed among dispersed teams and units of MNCs. Yet, research shows that successful MNCs today seek out, meld and leverage knowledge and learning practices on a global basis, moving from one-directional transfer strategies to multidirectional and highly reciprocal modes of joint learning and innovation engaging employees at many levels in innovative activities. Accordingly, and in contrast to traditional hierarchical models of MNC organization (e.g. Egelhoff, 1991; Vernon, 1966), recent studies suggest that the competitive advantage of MNCs increasingly derives from their potential to innovate in lateral collaborative networks (involving a multitude of organizational members) through joint learning activities based on distributed and highly multidirectional ways of transferring, sharing and recombining knowledge and practices previously compartmentalized around the world (Ghoshal et al., 1994; Nobel and Birkinshaw, 1998; Singh, 2005). However, limited research has been conducted on how MNCs create and organize conditions for joint learning to facilitate such practice-based forms of innovation in and across dispersed units and teams. Scholarship on knowledge transfer and global innovation networks (Bartlett and Ghoshal, 1990; Chesbrough, 2003; Chesbrough et al., 2006) describes the potentially enormous gains of knowledge-sharing and day-to-day innovative activities (Foss, 2006; Foss and Pedersen, 2004) as well as numerous barriers to creating, transmitting, integrating and deploying innovation between organizational members in MNC knowledge-sharing networks (Belangér et al., 1999; Grandori and Kogut, 2002; Gupta and Govindarajan, 2000; Kogut and Zander, 1993). Literature also emphasizes that collaborative activities among employees, teams and units are crucial for MNCs' innovative capabilities (Persuad, 2005) because they magnify the ability to accumulate knowledge and create new innovations through interactive learning (Lundvall, 1992; Lundvall et al., 2009). Accordingly, studies show how MNCs are redefining their roles (Bartlett and Ghoshal, 1986; Herrigel, 2007), organizational structures (Hedlund, 1994; Lord and Ranft, 2000) and governance practices (Ghoshal and Bartlett, 1990; Kristensen and Zeitlin, 2005; Whitley, 2001) towards more horizontal and collaborative activities facilitating co-practices of innovative learning in MNCs (Frost and Zhou, 2005; Ghoshal et al., 1994). Although studies have addressed the mechanisms and management of these novel forms of distributed innovation (e.g. Frost and Zhou, 2005; Singh, 2005), they typically consider only the aggregated firm level, the role of management (Bartlett and Ghoshal, 1989) or R&D-based innovative forms such as patent citation evidence (Rosenkopf and Almeida, 2003), competence-exploiting (Cantwell, 1995; 2005; Cantwell and Zhang, 2009), or firm external innovative search strategies (Laursen and Salter, 2006). To date, therefore, we have a limited understanding of how such innovative co-creation practices are organized, distributed and managed among organizational members at the level of everyday work-organizing practices. Furthermore, we lack micro-founded explanations of how employees may in fact become involved in and act as 'drivers' of ongoing innovative search, as well as knowledge on how such innovative practices may flow across various organizational and institutional contexts.

While firms in the Western world are experimenting with novel forms of work organization and drifting away from previous forms based on Taylorist management or 'patriarchal' authority, the transformation in the Nordic countries, in particular, has been characterized by a bottom-up process leading to new forms of 'learning organizations' characterized by extensive autonomy on the part of employees and work teams (Kristensen, 2010; Lorenz and Valeyre, 2003). In earlier studies (Lotz, 2009; Kristensen *et al.*, 2011) we have empirically demonstrated how Danish firms have engaged in ongoing innovation practices as 'learning organizations' where employees having high discretion, who are often closely linked to customers and suppliers, co-create innovations via joint learning and ongoing search routines fostered by novel forms of team-based work organization and the use of vocational training institutions. In these forms of organization all groupings of employees are constantly engaged in adapting and changing their own work, the tasks of the team or the team's relations and roles in regard to the larger landscape of teams – both within the firm and in other firms. This change has gone hand-in-hand with a radical upskilling of all employee groupings. Accordingly, in a series of Danish firms, we have observed employees working in cost-reducing teams while simultaneously improving processes and work arrangements and taking part in various ad hoc teams experimenting with user and employee-driven innovation activities (Kristensen, 2010: 178). These observed shifts in work organization, therefore, serve as a rich case for exploring the concrete 'organizational moves' that have facilitated EDI at the level of everyday work practices within Danish firms. Informed by this empirical field, next, we provide stories of four distinct 'organizational moves' towards EDI.

Empirical stories on the 'organizational moves' towards EDI in and across teams

The four 'organizational moves' concern the development of (1) a tight integration of planning and execution, (2) work-organizing practices premised on various criss-crossing teams, (3) a role division within teams and (4) a second job (JOB2) arrangement inviting employees to temporarily swap jobs with colleagues across units and teams. We do not claim that these stories of four distinct 'moves' towards the development of work-organizing practices that facilitate dynamics of EDI offer a set of fixed answers of how to organize for, distribute and govern EDI. Neither do we claim that one can generalize on the basis of these four stories, which are intrinsically shaped by the particular organizational and institutional context in which they are embedded. In drawing on these stories the aim is simply to carve out some possible routes towards constructing novel work organizational forms that may support EDI within current organizational life. Before we present the four 'organizational moves', we first provide a short description of the two case companies on which our study is based (pseudonyms are used and the companies' products are referred to in general terms).

Tools Ltd is a company that produces industrial tools but sees its main business as offering customers production-optimizing consultancy services, tool management and maintenance, and education and training. The company has built up its international capacity since 1995 to become a small Danish multinational with subsidiaries and sales offices abroad. Today it employs 650 people, of whom approximately 450 are working in Denmark. It is fully owned by management and employees, and ownership includes 85 per cent of the employees. The employees are teamworking in very unconventional physical facilities. As 'roofed villages', all Tools Ltd companies are designed along the ideals of a village community, with production and sales/administration situated in the same location and with no walls separating departments. Research was conducted at the company's headquarters in Jutland, in which approximately 450 employees working in production, stores and administration are placed on the same floor in one open room. By continuously expanding its activities, Tools Ltd has transformed itself from a small local supplier into a service-oriented total supplier, operating on the global scene in a tight interplay with customers in need of high-quality tools.

Health Ltd is one of the world's leading providers of medical analysers. The particular research site for the study was Health Ltd's main production plant, a Danish site with more than 800 employees (approximately 450 of them being so-called unskilled) located at the company's headquarters. While the production of medical analysers and instruments is the core métier, the company also offers a wide range of, for example, liquids, samplers and services such as process analysis, IT systems, quality and technical support, and training. Hence, co-developing collaborative communities (the company uses the term 'partnerships') with colleagues, consumers and suppliers is characteristic of Health Ltd's experimentalist work environment.

Our orientation to data generation in these two companies has been exploratory, using qualitative methods informed by a focus on the everyday work practices, work experiences, innovative activities and lives of organizational members. The methods deployed included formal and informal interviewing of organizational members at all levels, observations shadowing managers and employees, pictograms of organizational members' collaborative communities (Lotz, 2009) and secondary data (Spradley, 1979; Stake, 1998; Van Maanen, 1988). In total 55 formal interviews plus various informal observations and conversations were carried out. Data were in this way gathered from multiple sources within each organization and validated through a triangulating process of data evaluation that, as Hargadon and Bechky put it, builds support for any findings from the convergence of multiple, independent observations (Hargadon and Bechky, 2006: 489). Analytically, we have used 'opinion condensing', as inspired by Steiner Kvale (1997), and a coding technique that draws on categorization methods used in grounded theory (Glaser and Strauss, 1967; Strauss and Corbin, 1994) to develop knowledge about experimentation with novel work organizational forms facilitating EDI. Both case companies have built up work-organizing practices characterized by a significant ability (1) to collaborate in and across dispersed units, teams, and professional groupings, transgressing traditional organizational divides, (2) to continuously reorganize and recombine organizational resources (both human and non-human) in innovative ways and (3) to distribute intelligence and authority throughout the organization. Overall, our study shows that one of the central mechanisms through which EDI (and other practice-based forms of innovation) emerges is through experimentation with such novel forms of highly collaborative team-based work organization. The four 'organizational moves' presented below serve to illustrate more concretely how these work-organizing practices paved the way for EDI within the context of Tools Ltd and Health Ltd.

Dismantling rigid divides between planning and execution

The first story is about the development of a unique integration of planning and execution in Tools Ltd. A tradition of co-ownership comprising 85 per cent of employees constitutes a very collaborative and involving organizational order throughout the company. When the share scheme was implemented in 1977 the company employed 38 people. Today, approximately 475 employees are co-owners, holding a significant amount of stocks (typically worth approximately Euro 100,000 each). The price of stocks, dividends, and so on is dependent on the company's revenue, earnings and equity. In line with prior research on the effects of employee ownership (Buchko, 1992; Pierce and Rodgers, 2004), typically, organizational members in Tools Ltd. emphasize how the arrangement facilitates a work environment characterized by a high degree of involvement, mutual responsibility, commitment and distributed authority. Their stories about work indicate that these features trigger aspirations for knowledge-sharing and joint search practices for continuous improvements as well as the distribution of accountability and engagement at all levels. During our field studies we observed that the ownership structure in particular made possible two sets of institutionalized practices that gave rise to everyday innovation activities among employees. The first is centred on integrating work processes of planning and execution; the second (which we describe in the next section) is a fluctuating organizing form based on various crisscrossing collaborative communities.

In Tools Ltd people have deliberately worked (and are still working) towards integrating organizational processes of planning and execution and integrating their customers and partners into collaborative search practices for ongoing (co-)improvements and innovative co-creations (of products, services and practices). The company's physical layout, designed as a 'roofed village' in which production and sales/administration are situated in the same location with no walls separating departments or units, very much underlines this integrative work style. Deliberately drawing on both sources of knowledge (planning and execution) provides the basis for more flexible processes of co-creation driven by joint knowledge-sharing and distributed searches for improvements. In this way, former linear and less flexible production processes have been substituted by processes of experimentation based on close interactions between development and fabrication. In the literature (Helper et al., 2000) it is frequently emphasized that engineering has moved from being sequentially to becoming simultaneously organized, involving many units from several independent organizations in concurrent processes of co-creation. In Tool Ltd – and other Danish firms – this transformation has also included production workers as stakeholders in simultaneous engineering. Much more tentative and iterative work practices have evolved from this form of functional integration. Both employees and managers in Tools Ltd emphasize how this work style results in better innovative solutions and performance results, and therefore they all have a common interest in engaging in, developing with, and learning from these collective interdisciplinary activities. Working together with other professions (production workers, engineers, technicians, sales representatives etc.) strengthens the ability to learn, relearn and take on new work roles. Such encounters help organizational members question habitual routines and enable them to better exploit and explore their skills and work practices in renewing ways. The fact that they are at the same time being bound to each other by common interest and a common purpose seems to facilitate ongoing innovative search practices. Moreover, through close collaborative interactions with customers, the company has developed search processes which transgress traditional organizational boundaries by integrating customers and other 'outside' partners in their everyday search for continuous improvements.

Moves towards an organizational infrastructure of various criss-crossing teams

The second empirical story is about the development of work-organizing practices premised on various criss-crossing teams. In Tools Ltd the work within and across teams is highly collaborative, creating a pattern of employee and customer-driven innovations. The entire formation of teams operates in very informal and fluid ways. In principle, all areas of the company may become involved in changing combinations, depending on particular customer needs. Each unit (e.g. regrinding of metal tools, the calibration centre, the sales or construction department) represents a form of *basic team* divided into smaller sub-teams depending on the units' actual work functions, operations and work tasks. Strong professional bonds and feelings of pride towards the units' specific work activities, competencies and performance results characterize life within different basic teams. Simultaneously, more provisional *ad hoc teams* are continuously constructed across functions and operating units depending on the competencies and resources required for a given task or project (of search). Hence, collaborative team communities of organizational members from different units and with different skills and competencies are assembled on an ongoing basis. This way of forming criss-crossing temporary teams not only unfolds locally (for example in Denmark), but is today connecting all the sites of Tools Ltd on a global scale and at numerous levels. In this way the organizing practice of the roofed village circumscribes a global search organization. These more temporary and fluctuating collective practices that criss-cross the organization's basic team communities spur ongoing search processes which involve significant collaborative dynamics premised on the exploitation and exploration of organizational resources (routines and habits, materials, ideas, services etc.). In this way individual organizational members learn, for instance, to reflect upon their own and others' habitual ways of working, share knowledge, take on new roles, enhance competencies and create novel working practices and careers. The company's criss-crossing team organization, based on ongoing distributed search practices with hardly any centralized hierarchical coordination, seems to a large extent to be monitored by collective processes of involvement and the experience of a common purpose. Clearly, the joint ownership structure enhances involvement and helps coordinate these processes, in that it leads organizational members to work and connect for a common purpose. Although unique, Tools Ltd's co-ownership model and fluctuating configuration of various criss-crossing teams depict some interesting contours of how to organize, distribute and govern everyday innovation activities among employees in and across different teams and groupings.

Institutionalizing a role division in teams

In Health Ltd, two different institutionalized sets of practices – or distinct 'organizational moves' – have been used to foster collaboration and ongoing search activities within and across units and teams. The first is the institutionalization of a role division within all teams. The second is the institutionalization of a 'JOB2' arrangement. Both sets of practices have been enabled by a tradition for co-management between top production managers and conveners / shop stewards. This implies that union representatives have been integrated into the management structure. The partnership has cultivated the company's ability to involve and engage organizational members in innovative search activities within and across diverse collaborative teams. Since team organization was introduced into the company several decades ago, it has continually changed, with the overall purpose of giving employees increasing influence over the development of work practices. Employees in Health Ltd highlight how the partnership between top managers and union representatives has helped to strengthen collaboration across teams, enhanced transparency, and enabled the delegation of authority to teams and a better flow of information and knowledge-sharing within and between units and teams.

As a way of developing and monitoring work roles that facilitate the involvement and continuous development of employees' ideas, skills and competences, the company has implemented a role matrix within each production team, which, as well as their operational roles, includes: quality control responsibility, coordinator, documentation responsibility, capacity responsibility, stock responsibility, technical responsibility, education and training responsibility, environment responsibility, and information and IT responsibility. This set of formal roles has been developed by union representatives in collaboration with managers and employees. To reach the highest degree of flexibility, more than one person within each team must, as a rule, be able to perform each role, and employees are encouraged to change roles. With explicit roles the division of responsibility becomes more transparent, and, both across teams and in relation to wider surroundings, formal roles facilitate coordination and decentralization of work and authority. Moreover, 'role masters' with similar responsibilities exchange experiences and ideas across teams, creating a dynamic of learning and innovating across teams. Given that the team members perform the different coordinating roles and often do it in turn, they are all actively engaged in and share responsibility for the tasks and obligations of the nine roles. Through this form of mutual collaboration the individual team member becomes better able to take on the roles of others, reflect on these and thus become able to contribute with new ideas and suggestions for improvements. Such institutionalized practices of role-taking and role-shifting seem to strengthen the 'everyday' innovation activities and ongoing learning processes within the organization.

In many ways these responsibilities for making improvements have increased the pressure for and the propensity for employees at all levels to become increasingly innovative, resulting in organizations that are constantly on the move. But it has also created an organization where no-one has a comprehensive view over progress being made and the potential for innovations to become useful in other teams, units and departments of the company. The type of systematic and continuous improvement based on root-cause analysis and documented inventories of solutions that is often associated with lean management (Helper *et al.*, 2000) is underdeveloped, partly reflecting that everything is being changed all the time.

The development of a 'JOB2' arrangement

The fourth story is about an 'organizational move' towards the implementation of a permanent second job (JOB2) arrangement, which invites employees to swap jobs with colleagues across units on a temporary basis in order to learn new skills and competencies that, of course, also increase organizational flexibility. The JOB2 arrangement implies that workers are given the opportunity to and are expected to shift jobs across teams or units. In fact, many have voluntarily been 'expatriated' to a JOB2 more than once. The arrangement allows new skills and competencies to be acquired and increases the functional and numeric flexibility within the organization. Over the years the JOB2 arrangement has been readjusted several times. A committee of union representatives, managers and employees is in charge of generally managing and adjusting the arrangement, but the practicalities of job shifts are self-organized by the teams. The JOB2 arrangement is used in the following situations: (1) moving and adjusting capacity, (2) employees' wishes for new challenges, (3) to ensure flexibility in relation to bottlenecks and key functions, and (4) currently to expand the ability to change (Source: material from the JOB2 committee). Today 30 per cent of the labour force moves around in JOB2 every day. Employees do it on a voluntary basis and management has no influence. The JOB2 arrangement increases exchange of resources (of, e.g., people, best/worst practices, ideas and knowledge) across units, and it helps to build up a 'collective conscience' and various collaborative bonds across units, teams and skills. In this way new competencies and roles are assembled and mixed on an ongoing basis. The union representatives list the following advantages of the JOB2 arrangement:

- Employee capacity is movable within 24 hours
- Employees move on a voluntary basis to different units or different shifts
- Maintain abilities to learn
- Are kept in the labour market in spite of changed job conditions
- Feel more secure about change
- Job security creates pleasure, motivation and readiness to change
- Conflicts among employees can be solved without dismissals
- Culture and traditions are constantly affected and experience is exchanged across the organization (Source: material from JOB2 committee)

Looking at the many advantages listed above, the JOB2 arrangement represents a unique 'organizing instrument' that not only facilitates the coordination of collaborative dynamics within and across teams, but also enhances the possibilities for joint learning and distributed intelligence (triggering ongoing reflexive experiences) in and throughout these various collaborative settings. Yet, again, these processes of joint learning happen through and by the movements of individuals, and what is being transferred and diffused is to a great extent accidental rather than systematic. This empirical story thus illuminates how the creation of such overlapping collaborative team practices allows different perspectives and work practices to be more easily shared and transformed into fresh innovative understandings of organizational products, services, practices and so on, but also leaves the organization as such without clear templates for securing the systematic flow of newly gained knowledge.

Taken together, we have identified four 'organizational moves' towards facilitating processes of EDI. These are the development of (1) a tight integration of planning and execution, (2) work-organizing practices premised on various criss-crossing teams, (3) a role division within teams and (4) a second job (JOB2) arrangement inviting employees to temporarily swap jobs with colleagues across units and teams. The four empirical stories indicate that experimentation with new forms of work organization offers some promising routes towards enabling employees at all levels to become the drivers of innovation. Our findings show that Tools Ltd and Health Ltd in particular have paved the way for such routes by establishing work roles and routine-breaking work practices in which organizational members must collaborate across traditional organizational divides, adopt multiple perspectives, accept new challenges, move among work roles and participate in various collaborative relationships to solve problems and engage in an ongoing search for innovation. In this way, our research provides empirical insights into how organizational members, through experimentation with new forms of work organization, become more involved in the everyday organization, distribution and mutual governance of practice-based innovation activities within the context of units and teams located in Denmark. In this way our study adds content to the concept of 'learning organization' by contributing empirical knowledge on the everyday work-organizing practices, work roles and employee-driven search practices that enable such organizational forms.

Compared with the movement towards high-performance work organizations, which has been inspired to a much higher degree by lean organizational principles initiated by the managerial apex in Anglo-Saxon countries, the four organizational moves described above are significant for the way they involve employees at all levels. Lorenz and Valeyre (2003) found that the Netherlands and Denmark, together with the Nordic countries, had made much more progress than other countries towards such 'learning organizations', and our investigations suggest that this is because the origin of the organizational moves was very different from that in Anglo-Saxon countries.

First, because of strong unions and an industrial relations legacy of negotiation among the social partners, shop stewards and convenors became very early partners in searching for ways to reform work organizations. In the Nordic countries, and in Denmark particularly, unions pushed their members to opt for further training, almost creating a pressure on firms to offer employees new and grander challenges. This greatly increased the competences on the factory floor, yet it took more than daring managers to let people grow with growing tasks. To have people shift between differing assignments means changes not only in the situation for the people who move, but also in operating conditions for the teams they leave for another assignment. This means that systems of re-numeration, benchmarking performance and so on must be changed accordingly if the system is not to lead to endemic conflicts. In most cases in Denmark this could not have happened without a constant refinement of the negotiating order and increasing involvement of shop stewards and conveners, or, alternatively, a system such as that in Tool Ltd, where the absence of divisions and partitioned measuring makes flexibility independent of the opportunism that could result from such arrangements.

In Denmark, highly developed further training institutions and a strong legacy of company-wide negotiations have made the described organizational moves possible. In our view, however, this is not enough to make them sustainable. The pressure on people to take on novel tasks and work with shifting deadlines and at different locations is very high. Without daycare for children, eldercare for the older generation and help in undergoing partial shifts in profession during rapid organizational changes, people simply could not work under such a regime. The welfare state, with an abundance of social services, constitutes, so to speak, an important infrastructure for a work system of employee-driven innovation (Kristensen and Lilja, 2011).

Here let us pause for a moment. Has the Netherlands, for example, with a much less developed public service sector, not progressed along the same route of organizing work as the Nordic countries? It seems so, but there the difficult balance between work and family life has been addressed by a wide diffusion of part-time work. On the one hand this shows that such evolution can take place under different institutional conditions, but on the other it raises the question of whether firms have the ability to make role changes as fast as in the Nordic countries. In principle we think that it is possible to make novel use of the institutional landscape in any country to make moves in the direction of making more space for employee-driven innovation, but that each national business system will also create distinct limitations that might call for more encompassing reforms. During the 1990s Denmark overcame some of these limitations unintentionally when the government launched an active labour market policy that greatly expanded the groupings engaged in further training and thereby fostered a very inclusive growth pattern.

There is no doubt that these forms of work organization have been highly important for the prosperous global expansion of small and medium-sized enterprises in the Danish economy during the 1990s. They had the organizational capability for fast role change that was in such high demand during that period. But this also raises the crucial question of whether and how such work practices of joint search may be transferred, recombined and further developed across international borders. In the last section of this chapter, we discuss the need for addressing the international dimension of EDI in future research.

Future steps: organizing for EDI in a global perspective

Prior research (Kristensen and Lilja, 2011; Lotz, 2009) has shown how the development in work organization towards 'learning organizations' has enabled Danish firms to co-create new competitive ways of producing, organizing and innovating within and across various teams which allow them to collaborate and compete more successfully on a global (or transnational) scale. Yet, one of the central challenges for (Danish) firms that become multinationals is how to continue to build up 'learning organizations' and innovate when units and teams are separated by national borders and diverse organizational and institutional contexts so that domestic innovations spread abroad and foreign advances are integrated at home. It is much easier to imagine how lean forms of organizations can be copied and diffused by hierarchical intervention and design than it is to imagine that such learning organizations can spread to locations without a strong legacy of negotiation, further training institutions and public services.

From our study, we know that many Danish firms are trying to create 'learning organizations' across international borders. But is that even possible? How robust are these 'learning organizational' forms? How will such organization work in Denmark and in other countries, and what forms of divisions of labour are created among Danish and foreign subsidiaries? Are there different institutional resources that may complement foreign subsidiaries in playing distinct roles? Or can the Danish organizational forms be replicated and further developed across units and teams of multinationals operating in different regional and national fields? How can EDI be organized, distributed and governed globally? In order to achieve a better understanding of whether and how it is possible to organize for EDI across international borders, we need to address the challenges of facilitating and further developing work-organizing practices enabling EDI when firms expand abroad. Future research must therefore approach EDI globally by exploring whether and how it is possible to create experimental work practices that replicate the pattern of, for instance, Danish affiliates in different contexts where they establish foreign affiliates (e.g. in China, Brazil or India). Similar studies should be made in other countries. In addition, the findings presented in this chapter need to be further qualified and expanded by asking questions such as (1) what organizational practices of EDI can be 'offshored' (i.e. organized and distributed) from Danish affiliates to their foreign affiliates and (2) what managerial templates (i.e. governance principles) are engaged to ensure and further develop work-organizing practices facilitating EDI across local and global teams of MNCs operating in diverse organizational and institutional landscapes. In order to do so, future research into the everyday encounters and micro-dynamics of organizational members' search activities and innovative (inter)actions across geographically distributed teams is required.

Researching how Nordic multinationals transfer their organizational forms and practices is also an approach to understand and assess the possibilities for EDI in different institutional settings, and thus could become a way to diagnose needs for reform in other countries, which have only taken small incremental steps in this direction.

References

Bartlett, C.A. and Ghoshal, S. (1989). *Managing Across Borders: The Transnational Solution*. London: Century Business.
Bartlett, C.A. and Ghoshal, S. (1990). 'Managing innovations in the transnational corporation'. In Bartlett, C.A., Doz, Y. and Hedlund, G. (Eds) *Managing the Global Firm*. London: Routledge.
Belangér, J., Berggren, C., Björkman, T. and Köhler, C. (Eds) (1999). *Being Local worldwide. ABB and The Challenge Of Global Management*. Ithaca, New York and London: Cornell University Press.
Birkinshaw, J., Ghoshal, S., Markides, C., Stopford, J. and Yip, G. (2003). *The Future of the Multinational Company*. West Sussex: John Wiley and Sons Ltd.
Buchko, A.A. (1992). 'Employee ownership, attitudes, and turnover: an empirical assessment'. *Human Relations*, 45(7), 711–733.
Cantwell, J. (1995). 'The globalization of technology: what remains of the product cycle model?' *Cambridge Journal of Economics*, 19(1), 155–174.
Cantwell, J. (2005). 'MNE competence-creating subsidiary mandates'. *Strategic Management Journal*, 26(12), 1109–1128.
Cantwell, J. and Zhang, Y. (2009). 'The innovative multinational firm: the dispersion of creativity, and its implications for the firm and for world development'. In Collinson, S. and Morgan, G. (Eds) *Images of the Multinational Firm*. Chichester: John Wiley and Sons Ltd.
Carlsson, B. and Stankiewicz, R. (1991). 'On the nature, function and composition of technological systems'. *Evolutionary Economics*, 1(1991), 93–118.
Chesbrough, H. (2003). *Open Innovation. The New Imperative for Creating and Profiting from Technology*. Boston, Massachusetts: Harvard Business School Press.
Chesbrough, H., Vanhaverbeke, W. and West, J. (Eds) (2006). *Open Innovation. Researching a New Paradigm*. New York: Oxford University Press.
Christensen, J.L. (2003). 'Changes in Danish innovation policy – responses to the challenge of a dynamic business environment'. In Biegelbauer, P.S. and Borrás, S. (Eds) *Innovation Policies in Europe and the US – The New Agenda*. Hampshire: Ashgate.
Cooper, R.G. (1990). 'Stage–Gate systems: a new tool for managing new products'. *Business Horizons* (May–June), 44–54.
Dewey, J. (1938, 1997). *Experience and Education*. New York: Touchstone.

Dorf, M.C. and Sabel, C.F. (1998). 'A constitution of democratic experimentalism'. *Columbia Law Review*, 98, 267–473.
Doz, Y., Santos J. and Williamson, P. (2003). 'The metanational: the next step in the evolution of the multinational enterprise'. In Birkinshaw, J., Ghoshal, S., Markides, C., Stopford, J. and Yip, G. (Eds), *The Future of the Multinational Company*. West Sussex: John Wiley and Sons Ltd.
Edquist, C. (2005). 'Systems of innovation – perspectives and challenges'. In Fageberg, J., Mowery, D.C. and Nelson, R.R. (Eds) *The Oxford Handbook of Innovation*. Oxford: Oxford University Press.
Egelhoff, W.G. (1991). 'Information processing theory and the multinational enterprise'. *Journal of International Business Studies*, 22(3), 341–368.
Ellström, P. (2006). 'Praktikbaserade innovationsprocesser I communal verksamhet – ett Lärandeperspektiv (Practice based processes of innovation in a communality perspective)'. In Jonsson, L. (Ed.) *Kommunledning och samhällsutveckling (Management of municipalities and development of society)*. Lund: Studentlitteratur.
Fenwick, T. (2003). 'Innovation: examining workplace learning in new enterprises'. *Journal of Workplace Learning*, 15(3), 123–132.
Foss, N.J. (2006). 'Knowledge and organization in the theory of the multinational corporation: some foundational issues', *Journal of Management and Governance*, 10, 3–20.
Foss, N.J. and Pedersen, T. (2004). 'Organizational knowledge processses in the multinational corporation: an introduction', *Journal of International Business Studies*, 35, 340–349.
Frost, T.S. and Zhou, C. (2005). 'R&D co-practice and "reverse" knowledge integration in multinational firms'. *Journal of International Business Studies*, 36, 676–687.
Ghoshal, S., Korine, H. and Szulanski, G. (1994). 'Interunit communication in multinational corporations'. *Management Science*, 40, 96–110.
Glaser, G.G. and Strauss, A.L. (1967). *The Discovery of Grounded Theory: Strategies for Qualitative Research*. New Brunswick: Aldine Transaction.
Grandori, A. and Kogut, B. (2002). 'Dialogue on organization and knowledge'. *Organization Science*, 13, 224–232.
Gupta, A.K. and Govindarajan, V. (2000). 'Knowledge flows within multinational corporations'. *Strategic Management Journal*, 21, 473–496.
Hargadon, A.B. and Bechky, B.A. (2006). 'When collections of creatives become creative collectives: A field study of problem solving at work'. *Organization Science*, 17(4), 484–500.
Heckscher, C. and Adler, P.S. (Eds) (2006). *The Firm as a Collaborative Community. Reconstructing Trust in the Knowledge Economy*. Oxford: Oxford University Press.
Hedlund, G. (1994). 'A model of knowledge management and the N-form corporation'. *Strategic Management Journal*, 15, 73–90.
Hedlund, G. (1986). 'The hypermodern MNC – a heterarchy'. *Human Resource Management*, 25, 9–35.
Helper, S., MacDuffie, J P. and Sabel, C.F. (2000). 'Pragmatic collaborations: advancing knowledge while controlling opportunism'. *Industrial and Corporate Change*, 10(3), 443–483.
Herrigel, G. (2007). 'Roles and rules: ambiguity, experimentation and new forms of stakeholder-ism in Germany'. *Mimeographed manuscript*, University of Chicago, January 2007.
Herrigel, G. (2010). *Manufacturing Possibilities. Creative Action and Industrial Recomposition in the United States, Germany and Japan*. Oxford: Oxford University Press.

Jensen, M.B., Johnson, B., Lorenz, E. and Lundvall, B.Å. (2007). 'Forms of knowledge and modes of innovation'. *Research Policy*, 36 (2007), 680–693.

Jeppesen, L.B. (2005) 'User toolkits for innovation: consumers support each other'. *Journal of Product Innovation Management*, 22(4), 347–363.

Joas, H. (1996). *The Creativity of Action*. Cambridge: Polity Press.

Kogut, B. and Zander, U. (1993). 'Knowledge of the firm, combinative capabilities, and the replication of technology'. *Organization Science*, 3(3), 383–397.

Kristensen, P.H. (2010). 'Transformative dynamics of innovation and industry: new roles for employees'. *Transfer*, 16(2), 171–184.

Kristensen, P.H. and Lilja, K. (Eds) (2011). *Nordic Capitalisms and Globalization. New Forms of Economic Organizations and Welfare Institutions*. Oxford: Oxford University Press.

Kristensen, P.H. and Lotz, M. (2011). 'Taking teams seriously in the co-creation of firms and economic agency', accepted for *Organization Studies*, 32(11), 1465–1485.

Kristensen, P.H. and Zeitlin, J. (2005). *Local Players in Global Games. On the Strategic Constitution of a Multinational Corporation*. Oxford: Oxford University Press.

Kristensen, P.H., Lotz, M. and Rocha, R. (2011). Denmark: tailoring flexicurity for changing roles in global games. In Kristensen and Lilja (Eds) *Nordic Capitalisms and Globalization. New Forms of Economic Organizations and Welfare Institutions*. Oxford: Oxford University Press.

Kvale, S. (1997). *InterView. En introduction til det kvalitative forskningsinterview*. København: Hans Reitzels Forlag.

Laursen, K. and Salter, A. (2006). 'Open for innovation: the role of openness in explaining innovation performance among U.K. manufacturing firms'. *Strategic Management Journal*, 27, 131–150.

LO Rapport (2007). *Employee-Driven Innovation – a Trade Union Priority for Growth and Job Creation in a Globalised Economy*. Published by LO, The Danish Confederation of Trade Unions.

Lord, M.D. and Ranft, A.L. (2000). 'Organizational learning about new international markets: exploring the internal transfer of local market knowledge'. *Journal of International Business Studies*, 31, 573–589.

Lorenz, E. and Valeyre, A. (2003). 'Organizational change in Europe: national models or the diffusion of a new "one best way"?' *Paper prepared for the 15th Annual Meeting on Socio-Economics LEST*, Aix-en-Provence, 26–28 June.

Lotz, M. (2009). *The Business of Co-creation – And The Co-creation of Business*. PhD Series 15. Copenhagen Business School.

Lundvall, B.Å. (1992). *National Systems of Innovation: towards a Theory of Innovation and Interactive Learning*. London: Pinter Publishers.

Lundvall, B., Vang, J., Joseph, K.J. and Chaiminade, C. (2009). 'Innovation system research and developing countries'. In Lundvall, B.-Å., Van, J., Joseph, K.J. and Chaiminade, C. (Eds) *Handbook Of Innovation Systems and Developing Countries*. Cheltenham: Edward Elgar Publishing Ltd.

March, J. (1991). 'Exploration and exploitation in organizational learning'. *Organization Science*, 2(1), 71–87.

Mead, G.H. (1934). *Mind, Self and Society*. New York: The University of Chicago Press.

Nobel, R. and Birkinshaw, J. (1998). 'Innovation in multinational corporations: control and communication patterns in international R&D operations'. *Strategic Management Journal*, 19, 479–496.

Persuad, A. (2005). 'Enhancing synergistic innovative capability in multinational corporations: An empirical investigation'. *The Journal of Product Innovation Management*, 22, 412–429.

Pianta, M. (2005). 'Innovation and employment'. In Fageberg, J., Mowery, D.C. and Nelson, R.R. (Eds) *The Oxford Handbook of Innovation*. Oxford: Oxford University Press. pp. 568–598.
Peirce, C.S. (1877). 'The fixation of belief'. *Popular Science Monthly*, 12, 1–15.
Peirce, C.S. (1992). 'The first rule of logic'. In Ketner, K.L. (Ed.) *Reasoning and the Logic of Things*. Cambridge: Harvard University Press.
Pierce, J.L. and Rodgers, L. (2004). 'The psychology of ownership and worker-owner productivity'. *Group and Organization Management*, 29(5), 588–631.
Rosenkopf, L. and Almeida, P. (2003). 'Overcoming local search through alliances and mobility', *Management Science*, 49(6), 751–766.
Sabel, C.F. (2007). 'A real-time revolution in routines'. In Adler, P.S. and Heckscher, C. (Eds) *The Firm as a Collaborative Community. Reconstructing Trust in the Knowledge Economy*. Oxford: Oxford University Press.
Sabel, C.F., Gilson, R.J. and Scott, R. (2009). 'Contracting for innovation: vertical disintegration and interfirm collaboration'. *Columbia Law Review*, 109(3).
Schumpeter, J. (1934). *The Theory of Economic Development*. Harvard Scrip Magazine, January 1999: Scrip's review of 1998.
Singh, J. (2005). 'Collaborative networks as determinants of knowledge diffusion patterns'. *Management Science*, 51, 756–770.
Spradley, J.P. (1979). *The Ethnographic Interview*. Wadsworth Group. New York: Holt, Rinehart and Winston.
Stake, R.E. (1998). 'Case studies'. In Denzin, N. and Lincoln, Y. (Eds) *Strategies of Qualitative Inquiry*. Thousand Oaks, CA: Sage Publications. pp. 86–109.
Stark, D. (2009). *The Sense of Dissonance. Accounts of Worth in Economic Life*. Princeton University Press.
Stark, D. and Girard, M. (2002). 'Distributing intelligence and organizing diversity in new media projects'. *Environment and Planning* A, 34(11), 1927–1949.
Strauss, A. and Corbin, J. (1994). 'Grounded theory methodology – an overview'. In Denzin, N.K. and Lincoln, Y.S. (Eds) *Handbook of Qualitative Research*. London: Sage Publications. pp. 1–18.
TemaNord (2006). *Understanding User-Driven Innovation*. Nordic Council of Ministers, Copenhagen.
Trott, P. (1998). *Innovation Management and New Product Development*. Financial Times, Pitman Publishing.
Van Maanen, J. (1988). *Tales of the Field*. Chicago: The University of Chicago Press.
Vernon, P. (1966). 'International investment and international trade in the product cycle'. *Quarterly Journal of Economics*, 81, 190–207.
Von Hippel, E. (1988). *The Sources of Innovation*. New York: Oxford University Press.
Weick, K. (1979). *The Social Psychology of Organizing*. New York: Random House.
Wenger, E. (1998). *Communities of Practice Learning, Meaning, and Identity*. Cambridge: Cambridge University Press.
Whitley, R. (2001). 'How and why are international firms different? the consequences of cross-border managerial coordination for firm characteristics and behaviour'. In Morgan, G., Kristensen, P.H. and Whitley, R. (Eds) *The Multinational Firm. Organizing Across Institutional and National Divides*. Oxford: Oxford University Press.
Whitley, R. (2006). 'Innovation systems and institutional regimes: the construction of different types of national, sectoral and transnational innovation systems'. In Lorenz, E. and Lundvall, B.Å. (Eds) *How Europe's Economies Learn: Coordinating Competing Models* (pp. 343–380). Oxford: Oxford University Press.

10
Exploring the Employee-Driven Innovation Concept by Comparing 'Innovation Capability Management' Among German and Chinese Firms

Werner Fees and Amir H. Taherizadeh

In the exploratory study presented in this chapter, we aim at discovering the role of employees in relation to a firm's innovation capability management process. We use quantitative data gathered mainly through email survey from 163 German and 173 Chinese firms operating in both manufacturing and service sectors. All sampled firms have been examined based on five key dimensions of innovation capability management: strategy, innovative organization, external linkages, processes, and learning organization.

This study attempts to empirically find out the answers to the following questions. First, is there a significant difference between the two samples concerning their innovation capability management? Second, are these differences explainable from the employee-driven innovation perspective, that is, in light of the role of employees and their involvement in the innovation management process?

Overall results show that there is a significant difference between the German and Chinese firms with regard to their way of managing innovation capabilities in three dimensions – strategy, innovative organization and learning organization. These results are in favour of the German firms, that is, they have a leading position vis-à-vis the Chinese firms. The detailed results, further, show that the main difference between the two samples originates from German firms' ability to deploy the hidden or invisible innovation capability of their employees. Our results support and place more emphasis on the importance of employee-driven innovation, and suggest that European firms can sustain their competitive edge over emerging economies like China by investing more in their employees as their invaluable intangible resources, which cannot be easily imitated.

Introduction

In recent years, the race for gaining competitive advantage has grown into a never-ending global contest in which firms strive to get an edge over competitors and stay in business. Some have to strive to win back their market share, while some others try to fox their competitors by discovering totally new markets and tapping into unmet needs. Notwithstanding the different courses firms may take, the so-called 'international ranking' developed in recent years has seen the modern Western economies in the 'driver's seat', gaining competitive advantage mainly through their innovation power and innovation-oriented leadership skills. Similarly, according to Sternberg and Arndt (2001), innovation has been identified as a key driver of Europe's national and regional economies in the current phase of economic globalization. The logic is that relatively high-cost locations, such as Germany, Finland, Denmark and Switzerland, are globally sustainable only if they are able to develop and market highly innovative and competitive products and services with ever-reducing time to market periods (see also Sternberg and Arndt, 2001). But, recently, this leading position of European firms has been shadowed by the newly emerging economies such as China, where firms can manufacture products not only more cheaply but also under less restricted regulatory and supervisory conditions compared with the EU member countries, the USA and Canada. Historically, this has turned China into the workbench of the world, to the extent that many analysts believe that business in China has been all about volume and low labour cost! However, things have been changing recently.

The traditional view that a modern sense of innovation belongs to the West does not seem sustainable, since China's *economic transformation*, which has taken place in the last three decades, is now regarded as one of the most significant social changes in human history (Golley and Song, 2010). The country has undergone a successful transformation from a centrally planned, closed economy into one of the world's most dynamic and globally integrated market economies. China has already become aware of the importance of innovation in moving its economy forward and improving the nation's wellbeing. In 2006, China launched a national campaign to enhance its capability for innovation, and making strides towards increasing the country's innovation capability became a top priority for the government.[1] Likewise, in the spring of 2006, China held the first National Conference on Science and Technology (S&T) in the new century, the outcome of which was the outline of the *National Program for Long- and Medium-Term S&T Development (2006–2020)*. This major event led the central government to conceive the concepts of 'independent innovation' and 'innovative country' to become core strategies of China's development and the key to enhancing national strengths.[2] These initiatives led to the development of the 2020 vision for China to become an 'innovation-oriented' country; yet there would be a long way to go for China to arrive at the desired innovation status on a country-wide level.

In view of China's continued economic expansion and freer economy, European firms are facing the beginning of a new era of competition. The question is: when firms from countries like China, and likewise Brazil, India and Russia (the so-called BRICs), catch up with their European counterparts in terms of innovation, technology and market share dominance covering a wide range of products and services from low-tech to high-tech, what will remain as the next competitive edge in the hands of European firms?

This challenging question serves as the motivation behind conducting this study. We know that such questions are extremely challenging in nature and it is hard to find a cure-all answer, yet the upside is that they provide enough motivation for researchers from around the world to look at the issue from fresh and different interdisciplinary angles and provide their own contributions. In this study, we look at this important issue from a resource based view (RBV) of the firm that changes the focus from the external environment to viewing internal resources such as human potential (employees) and core competencies as key sources of competitive advantage, which may form new strategic options (Barney, 1991; Gagnon, 1999; Grant, 1991; Hamel and Prahalad, 1994; Wernerfelt, 1984). It may be very well the case that an emerging economy like China is endowed with a large volume of natural resources, capital and a vast workforce, but one should note that, in the absence of intangible qualities such as spirit and the right culture of innovation, tacit knowledge of experienced and well-trained employees, and values like integrity and empowerment, those quantitative measures will not count for much. In line with this argument, employees have been found to have hidden abilities for innovation (Cohen *et al.*, 1972; Ford, 2001), and these 'hidden abilities can be understood as an existing, albeit underutilized, resource' (Kesting and Ulhøi, 2010: 66), which can help European firms gain an edge.

Thus, viewing employees as important drivers of innovation has recently shifted the spotlight to the growth of Employee-Driven Innovation (EDI) concept among several European countries – Denmark, Finland, Italy and Germany, to name just a few (see, for example, Høyrup, 2010; Kristensen; 2010; Møller, 2010; Telljohann, 2010). In principle, EDI refers to 'the generation and implementation of significant new ideas, products, and processes originating from a single employee or the joint efforts of two or more employees who are not assigned to this task' (Kesting and Ulhøi, 2010: 66). Therefore, EDI focuses not only on experts but also on ordinary employees. EDI is still an under-researched concept (see, e.g., Høyrup, 2010) and little has been done to shed more light on how employees' potential can be realized (Kesting and Ulhøi, 2010). Therefore, in this study, we attempt to analyse and compare the innovation capability management of German and Chinese firms in the light of EDI. The following questions lead our empirical investigation:

1. Are there any significant differences between German and Chinese firms in managing their innovation capabilities?
2. In which key dimensions do these differences occur?
3. Are these differences explainable from the EDI perspective? What are the roles of employees in connection with these differences?

Study background

The term innovation comes from the Latin *innovare*, which means 'to make something new' (Tidd et al., 2001). The element of 'newness' can be found in various definitions of innovation which have been introduced into the literature. For example, Thompson (1965) defined innovation as the generation, acceptance and implementation of new ideas, processes, products or services for the first time within an organizational setting. Similarly, Damanpour (1991: 556) states that 'the adoption of innovations is conceived to encompass the generation, development and implementation of new ideas or behaviours.' Tidd et al. (2005: 66) also define innovation as 'a process of turning opportunity into new ideas and of putting these into widely used practice'. Overall, as definitions may differ slightly from one another, there seems to be a common understanding that innovation plays a pivotal role in the entrepreneurial process of wealth creation and helps firms play a dominant role in shaping the future of their industries. Therefore, in a borderless world where boundaries are no longer relevant, where accessibility to knowledge, expertise and technology is far advanced, innovation has become a necessity rather than a luxury (Kaplan and Warren, 2007).

However, knowing the importance and attributes of innovation is not enough for a firm to innovate. Firms' leaders need to know precisely what drives innovation to tailor their strategies towards fostering those drivers. Earlier work on innovation has extensively and intensively focused on R&D (Hatchuel et al., 2006; Lahiri, 2010), technology transfer (Howells, 1996; Marcotte and Niosi, 2005; Mowery and Oxley, 1995), market and consumers as main sources or drivers of innovation (see, for example, Kok and Biemans, 2009; Love and Roper, 1999; Lukas and Ferrell, 2000; Shefer and Frenkel, 2005). Yet, 'employee-driven innovation' (EDI) as a 'new form of innovation' (Høyrup, 2010) which takes 'ordinary' employees as an important driver of innovation has received little and/or secondary attention thus far. As stated by Høyrup (2010), expertise, experience, ideas, creativity and skills of the firm's employees are among the most important drivers of innovation. This perspective brings the employees and EDI to the forefront of innovation research in the European context. In simple terms, EDI refers to 'the generation and implementation of significant new ideas, products, and processes originating from a single employee or the joint efforts of two or more employees who are not assigned to this task' (Kesting and Ulhøi, 2010: 66). Similarly, it refers to the systematic nature of employees' contribution to the process of innovation (LO, 2007, as cited in Teglborg-Lefevre, 2010). In essence, the concept of EDI attempts to highlight that innovations can also come from 'ordinary' or lower-echelon employees, 'from shop-floor workers and professionals to middle managers across the boundaries of existing departments and professions' (Kesting and Ulhøi, 2010: 66).

In the past, EDI has also been the subject of investigation in proximity to or relationship with other concepts. For example, Tidd and Bessant (2009) and Tidd et al. (1997) researched EDI in the light of high-involvement innovation. They provided a comprehensive overview of the literature on the 'innovative organization' from which the role of employees in relation to the firm's vision, leadership, effective teamwork, and customer focus could be understood (as cited in Nijhof et al., 2002). Although Tidd et al. (1997) investigated the role of employees (indirectly/directly) under five key dimensions of innovation capability management, there were also others (see Imai, 1987) who undertook earlier research on the participation of employees in continuous improvement activities rather than in more radical innovations. Nijhof et al. (2002) also investigated the use of employees exempted from other tasks on innovative work.

As mentioned earlier, the concept of EDI is underpinned by the assumption that employees have hidden abilities for innovation and are considered among firms' innovation capabilities. This concept has been more recently studied in a broader context such as *non-R&D innovation* (see UNU-MERIT, 2008, as cited in Høyrup, 2010; Aho, 2005). One of the major theories supporting EDI is the resource-based view (RBV) of the firm which conceptualizes the enterprise as a 'bundle of unique resources' (Penrose, 1959). Penrose asserts that the growth of the firm is both facilitated and limited by management search for the best usage of available resources. These include tangible and intangible, human and non-human resources that are possessed or controlled by the firm and that permit it to devise and apply value-enhancing strategies (Barney, 1991; Wernerfelt, 1984). Therefore, assets, capabilities, processes, attributes, knowledge and know-how that are possessed by a firm and that can be used to formulate and implement competitive strategies are all inherently valuable resources of a firm. In the RBV view, employees at all levels of the organization are perceived as 'innovation capital' or 'innovation assets'– terms also supported by Kesting and Ulhøi (2010).

Moreover, the resource-based perspective invites consideration of managerial strategies for developing new capabilities (Wernerfelt, 1984). In fact, there is an issue of control over these scarce resources which are considered as the source of a firm's economic profits (Teece et al., 1997). Therefore, issues such as 'skill acquisition', 'the management of knowledge and know-how' (Shuen, 1994) and learning become fundamental strategic issues (Teece et al., 1997). Accordingly, strategic management researchers such as Teece et al. (1997: 514–515) believe that the greatest potential for contributions to strategy comes from this dimension, which encompasses skill acquisition, learning and accumulation of organizational and intangible or 'invisible' assets (Itami and Roehl, 1987). In this view, learning, innovation and competitive advantage are closely interlinked concepts which should be studied collectively.

Learning is a fuzzy construct which is multifaceted, multidimensional and not easy to operationalize; however, it is increasingly perceived as a key element in developing and sustaining a firm's competitive advantage (see Armstrong and Foley, 2003; Baldwin et al., 1997; DeGeus, 1988; Goh and Richards, 1997; Liedtka, 1996). Schein (1997) asserts that we do not have a clear understanding of the words 'organization' and 'learning'. However, many authors (Argyris and Schön, 1978; Grant, 1996; Huber, 1991; Spender, 1996) believe that *an organization learns through its individuals*. Maybe the following excerpt from Kim (1993: 44) will clarify this further:

> Imagine an organization in which all the physical records disintegrate overnight. Suddenly, there are no reports, no computer files, no employee record sheets, no operating manuals, no calendars, – all that remain are the people, buildings, capital equipment, raw materials, and inventory. Now imagine an organization where all the people simply quit showing up for work. New people, who are similar in many ways to the former workers but have no familiarity with that particular organization, come to work instead. Which of these two organizations will be easier to rebuild to its former status?

In Kim's view, organizational learning can be thought as a metaphor derived from his understanding of individual learning because, ultimately, organizations learn through their individual members and this shifts the focus more and more onto employees, their learning, and the outcome of their learning: innovations. Therin (2010) posits that 'learning leads to newness, and thus to innovation.' Thus, innovative firms place an emphasis on enhancing their employees' learning opportunities and they harvest the results through employee-driven innovations. We believe that EDI lies very much at the heart of the concept of *learning organization* where 'people continually expand by their capacity to create the results they truly desire, where new and expansive patterns of thinking are nurtured, where collective aspiration is set free, and where people are continually learning how to learn together' (Senge, 1990: 3).

On the relationship between organizational learning and successful innovations, we should note that the former does not necessarily cause the latter, as innovation and innovation's success are two different phenomena. A successful learning organization leads to the capacity to innovate (Burns and Stalker, 1961), which is the ability of the organization to adopt or implement new ideas, processes or products successfully (Hurley and Hult, 1998, as cited in Therin, 2010).

As a case in point, Taherizadeh (2010) shows that *employee competence*, which involves the elements of learning organization, is one of the key drivers of innovation among Malaysian firms and contributes significantly to firms' innovation performance. However, in the study conducted by Terziovski and Samson (2007) among Australian firms, 'people competency' does not turn out to be one of the best drivers of innovative organizations; *committed leadership* and a highly developed *innovation strategy* are found to be of more importance.

Although work has been done on the role of employees in relation to firms' innovation capability development and innovation performance, there is a lack of solid empirical evidence with converging results. In fact, Kesting and Ulhøi (2010: 67) asserted that 'research on EDI is still in its infancy'.

In this chapter, we do not 'directly' study the concept of learning organization or employees, but we attempt to empirically investigate and compare the process of innovation capability management between German and Chinese firms by analysing its five key dimensions: strategy, innovative organization, external linkages, processes, and learning organization. Based on the results under each dimension, explanations are made from the theoretical perspective of EDI.

Theoretical model and methodology

Model

In this study, we have adopted and slightly modified the innovation research model originally developed by Tidd *et al.* (2005). Figure 10.1 depicts the model of this study.

The strength of this model (Figure 10.1) is that it provides an *integrative approach* to the management of innovation, considering its market, technological, and organizational aspects. The idea is not only to focus on one dimension of innovation, such as R&D-driven or market-driven innovation, or supportive organizational structures, but to adopt an integrated approach which does not favour 'one best way' solution and seeks the interrelationships which exist among the various dimensions. Therefore, the process of innovation management is seen as being essentially generic, although internal and external factors will influence its optimum shape (Tidd et al., 2005). In line with this view, a successful innovation:

is strategy-based,
depends on effective internal and external linkages,
requires enabling mechanisms for making change happen,
only happens within a supportive organizational context,
requires an organizational learning process for building and integrating key behaviours into effective routines.

In essence, rather than concentrating on improving the performances within selected dimensions of innovation management, the aim for any firm should be to achieve a good all-round performance in all of these dimensions, that is, an integrated management of innovation. And employees play a pivotal role in each and every dimension of this multifaceted model.

Method and measurement instrument

This study employs a survey research method using the measurement instrument called *Innovation Capability Measurement Instrument* (ICMI). The items of ICMI are mainly extracted from the book *Managing innovation – Integrating technological, market and organizational change*. 3rd ed. (Tidd et al., 2005); however, based on the relevant innovation management literature, the authors have made the necessary amendments and modifications to accommodate the peculiarities of the contexts under investigation – China and Germany.

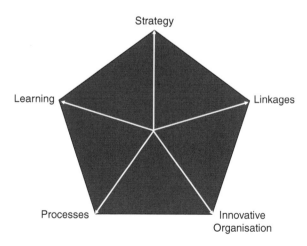

Figure 10.1 Key areas for successful innovation management
Source: Tidd et al., 2005: 568.

ICMI comprises 31 items encompassing the five key dimensions of our model: 1) Strategy, 2) Linkages, 3) Processes, 4) Innovative organization, and 5) Learning organization (Figure 10.1). Eight of the items are related to the company background (company size, business section, type of ownership and so on), and the remaining 23 items are built around our model's five key dimensions. The items are predominantly multiple choice questions allowing only one appropriate answer; however, there are some questions designed to elicit multiple responses. It is important to note that the results of this study should be viewed merely as the 'tip of the iceberg'; they only provide an exploratory indication of how innovation and employee involvement are being managed in the sampled firms.

Prior to administration, ICMI was translated into German and Chinese and checked for accuracy in line with the conventional backward translation method.

Respondents

In this study the unit of analysis was 'organization' and the target respondents were mainly senior managers and executives. Often, top managers were not very cooperative due to time constraints or simply lack of motivation; in such case, they were asked to nominate a senior employee who was very familiar with the firm and possessed sufficient knowledge to address ICMI. Moreover, students from executive MBA programs who held senior managerial positions were among the target respondents.

We made the effort to largely adopt a random sampling design, which tends to permit more generalizability power due to reduction of the bias involved in the process of data collection. However, a marginal part of our data (mostly from China) were collected through a critical case sampling (purposeful sampling) technique. We had to use some professional contacts with the businesses to obtain the required sample size due to the influence of *Guanxi* (关系) in the Chinese context. Apart from this marginal deviation, the majority of firms were randomly selected and contacted from major business websites and professional networks such as 'Thomas.net'[3] and 'Xing'[4] as well as associations such as 'the Chamber of Commerce and Industry of Central Franconia (Mittelfranken)'.

In China, sampled firms were mainly drawn from the Beijing, Hangzhou, Suzhou and Xian areas. Beijing was selected because its capacity and achievements in science and technology (S&T) are much stronger than other regions in China (Yam *et al.*, 2004). Suzhou and Hangzhou were chosen for their business–economic and industrial potential. Moreover, Suzhou is important due to its Suzhou Industrial Park (SIP), the largest economic and technological cooperation project between the Chinese and Singapore governments, as well as Suzhou Dushu Lake Higher Education Town (HET).

In addition, because of the proximity of Suzhou and Hangzhou to Shanghai, many businesses and industrial and manufacturing firms are

Table 10.1 Summary of the data collection

Item	Germany	China
Total number of target respondents approached	1114	350
Total number of valid responses received	163	173
Response rate (%)	14.63	49.42

clustered in these two cities. Therefore, much more emphasis has been placed on these two cities, to the extent that almost 90 per cent of responses came from these two regions.

Overall, we managed to collect data on 336 firms: 173 valid responses from China and 163 from Germany, representing the response rates of 49.4 per cent and 14.63 per cent respectively. Table 10.1 shows a summary of the data collection.

Data analysis technique

The data analysis is based on qualitative categorical as well as quantitative data. Simple statistical techniques (percentages) are used to analyse the gathered data. However, to have a condensed and objective analysis of the results, the qualitative data have been transformed into quantitative data by assigning a point scale system. In the following sections the results are presented in a tabulated form. It should be mentioned that the missing values are not explicitly shown, meaning that sometimes the figures do not add up to 100 per cent, and in some cases, where the respondents were allowed to choose multiple answers, the sum of the total percentages may exceed 100 per cent.

Descriptive results

Table 10.2 shows the descriptive characteristics of the participating firms. These characteristics include firm sizes (based on the total number of employees), firms' operating business sectors and ownership structures.

As shown in Table 10.2, the majority of the firms participating in our study are small and medium-sized enterprises (Germany: 85.8 per cent and China: 84.4 per cent). This classification is based on size definitions introduced by OECD;[5] Institut für Mittelstandsforschung Bonn (IfM)[6] and Hall's study (2007).

In Germany, a firm which has fewer than 500 employees is classified as an SME. Therefore, 85.8 per cent of the sampled German firms are SMEs. However, in the case of China, it is quite complicated to define an SME by Western standards, as SME definition can include relatively larger firms.[7] For example, in China, an industrial SME is a firm which has fewer than

2,000 employees. Therefore, it is natural to see relatively larger SMEs in China than in Germany.

In view of this, as the majority of the Chinese participants were from the industrial business sector (63.0 per cent), it makes sense to set 2,000 as a cut-off point to define SME; therefore SMEs represent 84.4 per cent of the sampled Chinese firms.

Table 10.2 also shows that the majority of sampled firms in China and Germany are from the industrial sector (China: 63.0 per cent and Germany: 47.2 per cent respectively). Firms from the service sector account for 20.8 per cent of our Chinese sample and 28.8 per cent of the German sample. In the German sample, 23.3 per cent of the firms represent the trade sector while only 11 per cent of the sampled Chinese firms come from this sector. The representation of banking and insurance sectors in both samples is too small to be significant.

Based on Table 10.2, it can be seen that the majority of the firms are nationally or locally owned (China: 74.0 per cent and Germany: 95.8 per cent). Therefore, the sample clearly reflects the ownership peculiarities of each nation state. Only a marginal percentage of our sampled firms have foreign owners (China: 14.5 per cent and Germany: 3.1 per cent) or have a mixed ownership structure (China: 11.6 per cent and Germany: 0.6 per cent).

In sum, our data largely represent SMEs of industrial sectors which have homogeneous local ownership.

Table 10.2 Descriptive data

Item	Germany (%)	China (%)
Number of employees		
10–50	46.6	19.7
51–200	28.2	32.9
201–500	11.0	16.2
501–1,000	4.9	8.1
1,001–2,000	1.2	7.5
2,001 and above	8.0	15.0
Business sector		
Industry	47.2	63.0
Service	28.8	20.8
Trade	23.3	11.0
Banking/Insurance	0.6	4.0
Ownership structure		
100% Local	95.8	74.0
100% Foreign-Owned	3.1	14.5
Joint Venture (Mixed)	0.6	11.6

Empirical findings and discussions

Aggregate results

This section provides the aggregate results of innovation management on five key dimensions by the German and Chinese firms (see Table 10.3 and Figure 10.2). A firm's performance on each dimension of innovation capability is measured on a scale ranging from 0 to 10, where '0' represents the absolute non-existence of the right capability on the respective dimension and '10' shows the full or maximum existence of the right capability on each dimension. In a practice, real-world scenario, no firm can score an absolute 0 or 10 and its score will fall somewhere in between. However, the higher the score, the more capable the firm is in managing its innovation management process; and the firm will be expected to perform better in comparison to its competitors in the same industry.

As shown by Table 10.3 and Figure 10.2, the German firms have a significantly better performance than their Chinese counterparts on three key dimensions of innovation management. These are, in descending order: (1) Innovative organization; (2) Learning organization; and (3) Strategy.

Regarding the two remaining dimensions, External Linkages and Processes, both groups of sampled firms show almost the same performance. However, there are marginal differences between the two samples which favour the German firms or put them in a better position. Yet, since these differences are not statistically significant, we think they might be due to random variations and therefore it is not suitable to discuss them in an aggregate form. In addition to these aggregate results, a new variable called *'Total Innovation Capability'* is created to determine which sample group, overall, takes the leading position in being more capable in managing innovation. This particular variable is the result of the total sum of the score points of all innovation dimensions. Therefore, theoretically, the minimum would be 0 and the maximum would be 50, with 0 representing lack of innovation management capability and 50 the full capability respectively. Table 10.4 shows the results.

Table 10.3 Average performance on each innovation dimension within German and Chinese firms

Dimension	Germany	China	Δ	Significance
Innovative organization	5.2	2.9	2.3	***
Learning organization	5.4	3.6	1.8	***
Strategy	3.9	3.6	0.3	*
External linkages	4.2	4.1	0.1	N.S.
Processes	3.1	3.0	0.1	N.S.

Notes: ***: Significance level is 0.001; *: Significance level is 0.1.
N.S.: Not significant.

As shown in Table 10.4, the German firms score, on average, significantly higher than the Chinese firms. Therefore, it can be concluded that in our sample, overall, the German firms are equipped with a more developed innovation management capability than their Chinese counterparts.

Based on the results presented in this section, we can answer the first two questions posed at the beginning of this chapter. Question 1: Is there a significant difference between German and Chinese firms concerning their *innovation capability management*? The answer is 'yes' – the German firms outperform the Chinese firms concerning their innovation capability management. Question 2: Under which dimensions do these differences occur? The answer is under three key dimensions – innovative organization, learning organization and strategy. Moreover, based on the aggregate results, *we hypothesize that German firms perform better than Chinese firms owing to practising a higher level of employee involvement in different aspects of the innovation management process.* This hypothesis makes more

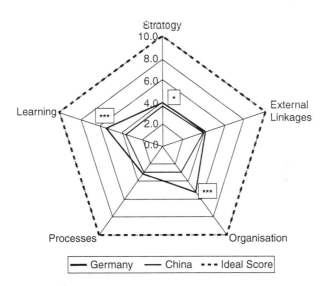

Figure 10.2 Average performance on each innovation dimension within German and Chinese firms

Table 10.4 Average total innovation capability in German and Chinese firms

New variable	Germany	China	Δ	Significance
Total innovation potential	21.8	17.2	4.6	***

Note: ***: Significance level is 0.001.

sense in the light of a quote from Tidd and Bessant (2009: 115): 'although each individual may only be able to develop limited, incremental innovations, the sum of these efforts can have far-reaching impacts.' In fact, the concept of high-involvement innovation is a reflection of what has been termed EDI (see, for example, Bessant, 2003; Høyrup, 2010; Tidd and Bessant, 2009). In the following section we attempt to provide a better understanding of the existing differences by presenting, analysing and discussing the detailed results under each innovation dimension in both samples.

Detailed results and discussion from the EDI perspective

In the current stream of research based on the European perspective on innovation, employees are considered very important (see Rocha, 2010; Kesting and Ulhøi, 2010; Kristensen, 2010; Møller, 2010; Tidd and Bessant, 2009, etc.) and have been found to be an effective (Høyrup, 2010) resource for a firm's innovation. However, the power of this essential element has often been neglected (Høyrup, 2010; Kesting and Ulhøi, 2010). Therefore, in this section, we report the detailed results on the five key dimensions of successful innovation management and discuss them in light of employees and the role they play with regard to enhancing innovation capability within an organization.

Following the concept of the integrated approach towards innovation management proposed by Tidd et al. (2005), employees' influence and participation play a pivotal role under each innovation dimension, and the major parts of this concept are built on the assumption that employee participation is essential for building the innovation capability of a firm. Further, this approach highlights employees' qualifications and continuous education/training as an invaluable contribution to the company's strategic orientation. The critical precondition for efficiently and effectively utilizing the employees' potential is for a firm to be equipped with a decentralized and democratic organizational structure which enables bottom-up and cross-functional communication, combined with a leadership style fostering delegation and empowerment.

Thus, this section aims to investigate how far differences in managing a firm's innovation capability are triggered by the firm's view of the employee-driven innovation concept. The results of this empirical study can be used to develop recommendations for the German and Chinese firms concerning how to improve their innovation capabilities, not only to maintain the strength of their status quo and overcome the weaknesses, but also to aim for a sustainable global competitive advantage by improving their innovation capability, more particularly, through investing in their employees. Table 10.5 shows an overview of the empirical results referring to the EDI-relevant variables with representative items.

Table 10.5 Detailed comparative analysis of five key dimensions of successful innovation management

Key innovation dimension	Germany (%)	China (%)	More towards nurturing EDI
1. Strategy			
– A clear strategy exists and is communicated and explained to the employees at different levels	42.3	29.5	Germany
– Strategic discussions are only held by top management	10.4	35.8	Germany
– Strategic discussions are held by top management, but the relevant employees are involved	57.1	25.4	Germany
2. Innovative organization			
– Clearly defined structure, particularly with overlapping responsibilities	64.4	26.0	Germany
– Important decisions are always taken by the CEO	42.9	60.7	Germany
– Decisions are made by appropriate employees without extensive degree of formality	14.7	0.6	Germany
– Communication within organization is typically top-down	20.2	60.7	Germany
– Communication within organization is typically bottom-up	8.0	22.0	China
– Communication within organization is totally open in all directions	69.9	9.2	Germany
3. Processes			
– There are processes in place to systematically search for new ideas	31.0	26.0	Germany
– Management seriously discusses employees' ideas	42.9	35.8	Germany
– We have organizational flexibility to quickly realize new ideas	57.1	17.3	Germany
– All employees are involved in the idea realization process	31.3	32.4	China/Germany
4. External linkages			
– Tight collaboration with suppliers	30.1	24.3	*
– Tight collaboration with customers	51.5	28.3	*
– Customers as main origin of ideas	75.5	40.5	*
– Tight relationship with scientific institutions	9.2	13.3	*

(continued)

Table 10.5 continued

Key innovation dimension	Germany (%)	China (%)	More towards nurturing EDI
– Regular linkages exist with external consultants	6.1	28.9	*
– The firm is a fixed part of a consultancy network	9.8	4.6	*
5. Learning organization			
– Majority of employees have no specific qualifications	8.0	16.2	Germany
– Only managers are well qualified	10.4	31.8	Germany
– Majority of employees are well qualified	80.4	39.9	Germany
– Further education once, twice or more often per year	37.4	19.1	Germany

*: these items are not directly linked to EDI but elicit the overall innovation capability on this dimension.

Strategy

Effective innovation management should be embedded in a systematic strategic management within a firm, that is, the firm's innovation activities are in line with its overall strategic objectives. Therefore, a direct link should exist between the overall strategic objectives and the innovation activities. According to Luecke (2009: 214), if creative people do not understand where the company is headed, they are likely to generate and pursue ideas that do not fit with current capabilities, that eat up resources, and that will eventually be rejected prior to commercialization. This will result in wasting the firm's financial resources and it dissipates energy. Therefore, in order to save resources (tangible and intangible), it is important to communicate the firm's strategy clearly across the organization and encourage employees to generate ideas within boundaries specified by the firm's strategy. This requires the involvement of broad parts of the organization in strategic discussions and will lead to very goal-oriented innovations with optimum use of resources (see also Kaplan and Norton, 1996).

As Table 10.5 shows, German firms more often have a clear strategy in place which is also communicated and explained to their employees, while this is not so often the case among the Chinese firms (Germany: 42.3 per cent vs. China: 29.5 per cent). In addition, the German firms tend to put more emphasis on employee involvement in their strategic discussions (Germany: 57.1 per cent vs. China: 25.4 per cent), while the Chinese firms prefer to restrict their strategic discussions exclusively to top management level (China: 35.8 per cent vs. Germany: 10.4 per cent).

Overall, as far as the strategy dimension is concerned, the German firms exercise a higher degree of employee involvement and highlight their employees' role in building a more competitive strategy. This approach is intended to incorporate the information and knowledge of the non-management-level employees (lower-echelon/ordinary employees) into higher-level (upper-echelon employees) organizational decision-making processes (Benson et al., 2006; Guthrie, 2001; Lawler, 1986). The fact that German firms take their employees more seriously and attempt to highlight their active presence in the strategic management process explains for the most part the significant differences which exist between German and Chinese firms' strategy dimension.

Innovative organization

> Innovation has nothing to do with how many R&D dollars you have. When Apple came up with the Mac, IBM was spending at least 100 times more on R&D. It's not about money. It's about the people you have, how you're led, and how much you get it. (Steve Jobs in *Fortune*, 9 November 1998)

Organizational issues such as structure, culture and environment, decision-making processes and communication patterns are typically in the focus of discussions in innovative firms. For example, Kanter et al. (1997) postulate several environmental factors which stifle innovation. These are dominance of restrictive vertical relationships; poor lateral communications; limited tools and resources; top-down dictates; formal, restricted vehicles for change; reinforcing a culture of inferiority which means 'the neighbour's grass is greener' or that innovations which come from outside are always better.

Past research has shown that firms' practices to involve employees in enhancing their knowledge, skills and abilities as well as their motivation to perform have been linked to positive organizational outcomes (Combs et al., 2006; Yalabik et al., 2008).Therefore, for an organization to be innovative, building a supportive and creative environment is a must. To realize this end, an organization needs to systematically develop organizational structures which reflect a positive culture and climate for innovations, allow a low degree of centralization and a high degree of employee involvement in innovation projects, set open communication policies and procedures, and capitalize on, especially, cross-functional teamwork activities.

As shown in Table 10.5, 64.4 per cent of the German firms have a clear organizational structure in place, compared with only 26.0 per cent of the Chinese firms. This means that the German firms have already institutionalized a clear division (and delegation) of tasks whereas the Chinese firms prefer a strict one-way decision line without allocation of responsibilities.

This interpretation is supported by the fact that in the Chinese firms the decision-making process is highly centralized and important decisions are always made by their CEOs (China: 60.7 per cent vs. Germany: 42.9 per cent), while the German firms have a certain degree of power delegation and employees are empowered to a certain extent to make decisions, or at least to participate in the decision-making process.

Although employees' nearly autonomous decision-making is not very high among German firms (14.7 per cent), it is still substantially superior to their Chinese counterparts (0.6 per cent), and this shows the high degree of trust in employees which exists in several German firms.

In the majority of the Chinese firms, 'communication' is typically top-down (60.7 per cent); that is, the upper echelon gives the orders and the lower-echelon follows. This is typically the case in only 20.2 per cent of the German firms. Of the Chinese firms, 22.0 per cent have a bottom-up communication mode, which seems markedly higher than their German counterparts (8.0 per cent). But, in terms of the possibility of totally open communication, meaning that there is no barrier to employees approaching their manager with an idea, a question, a concern, or simply to challenge something they do not agree with, German firms take the leading position (69.9 per cent), while fewer than 10 per cent of the Chinese firms practise such an approach to communication within their organizations.

Overall, decision-making is more centralized among Chinese firms and strictly restricted to the top management class, communication is mainly top-down and open communication hardly exists. Therefore, in such a bureaucratic climate, where employees hardly have room to express themselves or to be included in decision-making process, innovations are hard to come by. This lack of a positive and nurturing environment, which has turned the focus away from employees as a central pillar of innovation, accounts for the significant difference which remains between German and Chinese firms concerning the innovative organization dimension (see Table 10.3 and Figure 10.2).

Processes

Although some innovations are merely accidental and others may originate in a flash of inspiration, according to Peter Drucker, most of them are the result of a conscious and purposeful search for opportunities to solve problems or please customers (Drucker, 1985). Therefore, to make the innovation process more effective, organizational routines should be in place to systematically accelerate and facilitate this process from idea generation to commercialization. Even in the case of small firms which face constraints on gaining access to critical resources and capabilities for innovation (Hewitt-Dundas, 2006), a certain degree of formalization is necessary to ensure that ideas are handled diligently, employees are taken seriously, and management is seen as committed to innovation. Of course, a small company should not be 'over-formalized', losing the flexibility that is one of its core strengths; but even here some guidelines should be followed to make use of the advantages of an organized innovation process.

As shown in Table 10.5, 31.0 per cent of the German firms and 26.0 per cent of the Chinese firms demonstrate having systematic processes in place to search for new ideas. Of the German firms, 42.9 per cent mention that their top management seriously discusses their employees' ideas, as do 35.8 per cent of the Chinese firms. Both samples perform almost the same with regards to involving their employees in the idea realization process (Germany: 31.3 per cent and China: 32.4 per cent). The only notable difference between the samples comes from the organizational flexibility to realize new ideas quickly. Of the German firms, 57.1 per cent have organizational flexibility, while only 17.3 per cent of the Chinese firms are able to realize their innovation projects quickly. This indicates that the German firms have been more successful than the Chinese firms in overcoming bureaucratic barriers.

Overall, neither sample has outstanding performance with regards to having innovation processes in place, or demonstrates high employee involvement, and this is evident from their similar and insignificant performance on this dimension.

External linkages

A firm's internal competencies can be extended through horizontal or vertical linkages; and internal gaps (e.g. resources or know-how deficits) can be bridged through cooperating with partners. In managing a firm's innovation, linkages with customers, suppliers, scientists and competitors are essential to enhance the firm's innovation capability. In fact, through such linkages, know-how and resource gaps can be closed, customer orientation can be maximized and cross-company innovations can be realized. As Tidd et al. (2005) put it, it is now a 'co-operative federation of players!' In the same vein, Duysters et al. assert that 'alliances are no longer regarded as peripheral, but as a cornerstone of the firm's technological strategy' (1999, as cited in Nieto and Santamaria, 2010: 46). Furthermore, Nieto and Santamaria (2010) find that vertical collaboration with suppliers and clients has the greatest impact on Spanish manufacturing SMEs' innovativeness.

In view of the literature, the issue is not whether or not to collaborate with others, but to find out who spearheads this collaboration and to what extent the ordinary employees should be involved in these collaborations. If researchers and practitioners come to the conclusion that employees make a significant contribution to this networking, then maybe firms should consider re-engineering their organizational structure to place more emphasis on their employees and delegate more responsibilities to them, and, of course, hold them accountable accordingly.

In our study, however, the direct influence of employees on the efficiency and performance of external linkages has not been investigated. Therefore, in this dimension there is an avenue open for further, in-depth research to highlight employees' role in a firm's external linkages and the extent to which they contribute to innovations. Bearing this in mind, Table 10.5 presents the comparative results on this dimension.

As shown in Table 10.5, slightly more than half of our German sample (51.5 per cent) have tight collaboration with their customers, while only 28.3 per cent of the Chinese firms have established tight collaborations. Also, customers are used as the main source of new ideas more by the German firms than the Chinese firms: 75.5 per cent and 40.5 per cent respectively. Last but not least, the Chinese firms prefer linkages with external consultants (28.9 per cent) and scientific institutions (13.3 per cent) more than the German firms do (6.1 per cent and 9.2 per cent respectively). Notwithstanding these results, the role of employees in this dimension is under-researched, and further investigation of EDI in relation to a firm's external linkages is encouraged.

Learning organization

The learning organization is defined as 'a place where employees excel at creating, acquiring, and transferring knowledge' (Garvin *et al.*, 2008: 110). Pedler *et al.* (1991:1) stress that a learning organization 'facilitates the learning of all its members and continuously transforms itself'. Several authors emphasize generative learning that leads to innovation as a defining characteristic of the learning organization (Gardiner and Whiting, 1997; McGill *et al.*, 1992; Senge, 1990) and they view innovation as an important outcome and benefit of the learning organization (Porth *et al.*, 1999; Teare and Dealtry, 1998; as cited in Kontoghiorghes *et al.*, 2005). Therefore, innovation and learning walk the same path and one cannot study a firm's innovation capability without touching upon the concept of the learning organization, in the centre of which lies 'employee training and workplace learning'.

In view of this, it is highly important for firms to make the effort to create on-the-job learning opportunities for their employees so that they can better contribute to the firm's innovations. In our sample, as shown in Table 10.5, overall, the German firms have the attributes of a learning organization more often than the Chinese firms do. In our German sample, the majority of employees are well qualified (80.4 per cent); compared with the Chinese sample, more firms state that they invest in their employees' further education regularly (37.4 per cent); and just a minority of the firms state that only managers should be well qualified (10.4 per cent).

On the contrary, only 39.9 per cent of the Chinese firms state that the majority of their employees are qualified, 31.8 per cent of them mention that only top managers are well qualified, and only 19.1 per cent invest in their employees' further education.

These results support the overall significant difference which exists between the German and Chinese firms in the learning organization dimension (see Table 10.3). However, Germany's results should not be surprising, as lifelong learning is a core concept and a buzzword on European Union agendas and many efforts have been made in this direction.

The detailed analysis of the five key dimensions mentioned thus far provides insights into the way we look at the existing differences between German and Chinese firms' innovation capability management, and it provides an answer to the third question of this chapter. Not all differences can be explained by focusing only on employees and the roles they play in their organizations; many of them are attributed to the way firms perceive their employees' capability and capacity in contributing to firms' innovations. Issues such as involving employees in decision-making processes, clearly communicating the firm's strategic orientation concerning innovation strategy, establishing an open communication system, delegating authorities and, more importantly, providing necessary training, further education and workplace learning are among the main issues from an EDI perspective.

Limitations and avenues for further research

It is worthwhile to discuss the limitations of any study in order to help readers properly assess the validity and generalizability of the findings. In fact, limitations highlight the main barriers of the research and shed light on the weaknesses, which can be further eliminated or improved by future researchers who may wish to investigate the same topic and contribute to the relevant literature. This study is no exception and has four main limitations.

First and foremost, our sample size is not very large, and this does not permit us to use advanced statistical procedures. Therefore, making strong generalizations and extending the findings to all firms in each respective country would not be advisable. Second, the majority of our sampled firms are SMEs, and as a precaution it should be noted that the German SMEs are operating in a very established supportive national innovation system which is almost absent in China; therefore, the behaviour of the SMEs in two different contexts is quite different, and this may impact the interpretation of the results. Third, the data collected for the purpose of this study do not cover all geographical regions of the two countries under investigation. Therefore, the interpretations are limited and not relevant country-wide. Lastly, as we have not collected data from a specific industry, one cannot attribute the findings to a certain industry, and the interpretations should be made on a general basis.

The upside of the limitations is as follows. First, they reflect some realities regarding the nature of the research. For example, collecting a large sample size from managers in China or Germany representing a very specific industry seems to be a daunting task which may hinder researchers from even starting it. Second, limitations open windows of opportunity for further research. Since our findings are exploratory in nature, follow-up studies are encouraged to validate these results by exploiting a larger number of firms and running more rigorous and robust statistical procedures.

Moreover, future research is needed to delve further into employee-driven innovation by examining its direct effects on firm innovation performance across different regions in Europe. Further to this, qualitative research can be conducted not only to discover where the differences lie but also to explain why such differences exist.

Conclusion

In this paper we analysed, compared and discussed the German and Chinese firms' innovation capability management processes and made an effort to discuss their significant differences from the employee-driven innovation (EDI) perspective.

Our results show that: (1) the sampled German firms have a higher average total innovation capability compared with the Chinese firms; (2) the German firms have a significantly better performance than their Chinese counterparts on three key dimensions of innovation capability management. These are, in descending order: *innovative organization, learning organization* and *strategy*. Furthermore, the deeper analysis of the items under each of these dimensions reveals that these differences mainly exist due to high involvement of employees in the German firms.

In our sample, the Chinese firms are mainly characterized as being centralized organizations with a bureaucratic climate where employees are hardly given the opportunity to express themselves, to learn and to communicate freely in their organizations. Also, employee empowerment is not widely exercised in these firms.

Notwithstanding the existing differences and the superior innovation capability management of the German firms vis-à-vis the Chinese firms, it can be seen that the German firms are still not fully capable of optimally utilizing their innovation capability, including their employees' capability to drive innovation. This means that sooner or later German firms' superiority concerning their innovation edge, which has created the gap so far, may be overshadowed by emerging economic powers like China.

Therefore, these results can provide researchers with some preliminary insights into the topic at hand and benefit practitioners by emphasizing the role of employees in enhancing a firm's innovation capability. It is hoped that firms which lack innovation capabilities will try to revisit their view on the role of their employees in their innovation processes and put more emphasis on high employee involvement and EDI concepts.

Innovation Capability Management 207

Notes

1. 19 March 2006. China's innovation campaign: dos and don'ts. *China Daily*. Retrieved 25 April 2011, from http://www.chinadaily.com.cn/english/doc/2006-03/19/content_545827.htm
2. Chinese Academy of Science and Technology for Development. Retrieved 26 April 2011, from http://www.casted.org.cn/en/web.php?ChannelID=54
3. http://www.thomasnet.com/ (accessed during 2009 and 2010).
4. https://www.xing.com/ (accessed during 2009 and 2010).
5. OECD: http://stats.oecd.org/glossary/detail.asp?ID=3123 (accessed 15 May 2011).
6. Institut für Mittelstandsforschung (IfM) Bonn: http://www.ifm-bonn.org/ (accessed 20 January 2010)
7. SME Department of National Development and Reform Commission (NDRC) of P.R. China 2007-5-16, 'Interim Regulations on SME Categorizing Criteria'

References

Aho, E. (2005). *Creating an Innovative Europe*. Report of the group of independent experts on R&D and innovation appointed following the Hampton Court Summit.
Argyris, C. and Schön, D.A. (1978). *Organizational Learning: A Theory of Action Perspective*. Massachusetts: Addison-Wesley.
Armstrong, A. and Foley, P. (2003). 'Foundations of a learning organization: organization learning mechanisms'. *Learning Organization*, 10(2), 74–82.
Baldwin, T.T., Danielson, C. and Wiggenhorn, W. (1997). 'The evolution of learning strategies in organizations: from employee development to business redefinition'. *Academy of Management Executive*, 11(4), 47–57.
Barney, J.B. (1991). 'Firm resources and sustained competitive advantage'. *Journal of Management*, 17(1), 99–120.
Benson, G.S., Young, S.M. and Lawler III, E.E. (2006). 'High-involvement work practices and analysts' forecasts of corporate earnings'. *Human Resource Management*, 45, 519–537.
Bessant, J. (2003). *High-Involvement Innovation*. Chichester: John Wiley and Sons.
Burns, T. and Stalker, G.M. (1961). *The Management of Innovation*. London: Tavistock.
Cohen, M.D., March, J.G. and Olsen, J.P. (1972). 'A garbage can model of organizational choice'. *Administrative Science Quarterly*, 17(1), 1–25.
Combs, J., Liu, Y., Hall, A. and Ketchen, D. (2006). 'How much do high-performance work practices matter? A meta-analysis of their effects on organizational performance'. *Personnel Psychology*, 59, 501–528.
Damanpour, F. (1991). 'Organizational innovation: a meta-analysis of effects of determinants and moderators'. *Academy of Management Journal*, 34(3), 555–590.
DeGeus, A.P. (1988). 'Planning as learning'. *Harvard Business Review*, 66(2), 70–74.
Drucker, P.F. (1985). 'The discipline of innovation'. *Harvard Business Review*, May–June, 65–67.
Duysters, G., Kok, G. and Vaandrager, M. (1999). 'Crafting successful strategic technology partnerships'. *R&D Management*, 29(4), 343–351.
Ford, R.C. (2001). 'Cross-functional structures: a review and integration of matrix organization and project management'. *Journal of Management*, 18(2), 267–294.
Gagnon, S. (1999). 'Resource-based competition and the new operations strategy'. *International Journal of Operations and Production Management*, 19(2), 125–138.

Gardiner, P. and Whiting, P. (1997). 'Success factors in learning organizations: an empirical study'. *Industrial and Commercial Training*, 29(2), 41–48.
Garvin, D.A., Edmondson, A.C. and Gino, F. (2008). 'Is yours a learning organization?' *Harvard Business Review*, March, 109–116.
Goh, S.C. and Richards, G. (1997). 'Benchmarking the learning capability of organizations'. *European Management Journal*, 2(2), 575–583.
Golley, J. and Song, L. (2010). 'Chinese economic reform and development: achievements, emerging challenges and unfinished tasks'. In Garnaut, R., Golley, J. and Song, L. (Eds) *China: The Next Twenty Years of Reform and Development* (pp. 1–18). Australia: ANU E Press.
Grant, R.M. (1991). 'The resource-based theory of competitive advantage: implications for strategy formulation'. *California Management Review*, 33, 114–135.
Grant, R.M. (1996). 'Toward a knowledge-based theory of the firm'. *Strategic Management Journal*, 17, 109–122.
Guthrie, J.P. (2001). 'High-involvement work practices, turnover and productivity: evidence from New Zealand'. *Academy of Management Journal*, 44, 180–190.
Hall, C. (2007). 'When the dragon awakes: internationalisation of SMEs in China and implications for Europe'. *CESifo Forum* 8(2), 29–34.
Hamel, G. and Prahalad, C.K. (1994). *Competing for the Future*. Boston: Harvard Business School Press.
Hatchuel, A., Weil, B. and Le Masson, P. (Eds) (2006) *Les processus d'innovation*. Paris: Hermès.
Hewitt-Dundas, N. (2006). 'Resources and capability constraints to innovation in small and large plants'. *Small Business Economics*, 26, 257–277.
Howells, J. (1996). 'Tacit knowledge, innovation and technology transfer'. *Technology Analysis and Strategic Management*, 8(2), 91–106.
Høyrup, S. (2010). 'Employee-driven innovation and workplace learning: basic concepts, approaches and themes'. *Transfer European Review of Labour and Research*, 16(2), 143–154.
Huber, G. (1991). 'Organizational learning: the contributing processes and the literature'. *Organizational Science*, 2(1), 88–115.
Hurley, R.F. and Hult, G.T. (1998). 'Innovation, market orientation, and organizational learning: an integration and empirical examination'. *Journal of Marketing*, 62, 42–54.
Imai, K. (1987). *Kaizen*. New York: Random House.
Itami, H. and Roehl, T.W. (1987). *Mobilizing Invisible Assets*. Cambridge, Massachusetts: Harvard University Press.
Kanter, R.M., Kao, J. and Wiersema, F. (Eds) (1997). *Innovation: Breakthrough Thinking At 3M, DuPont, GE, Pfizer and Rubbermaid*. USA, New York: HarperCollins Publishers Inc.
Kaplan, M.J. and Warren, A.C. (2007). *Patterns of Entrepreneurship*. 2nd edn. USA: John Wiley and Sons, Inc.
Kaplan, R.S. and Norton, D.P. (1996). Using the balanced scorecard as a strategic management system. *Harvard Business Review*, 74(1), 75–85.
Kesting, P. and Ulhøi, J.P. (2010). 'Employee-driven innovation: extending the license to foster innovation'. *Management Decision*, 48(1), 65–84.
Kim, D.H. (1993). 'The link between individual and organizational learning'. *Sloan Management Review*, Fall, 37–50.
Kok, R.A.W. and Biemans, W.G. (2009). 'Creating a market-oriented product innovation process: a contingency approach'. *Technovation*, 29(8), 517–526.

This article is not included in your organization's subscription. However, you may be able to access this article under your organization's agreement with Elsevier.

Kontoghiorghes, C., Awbrey, S.M. and Feurig, P.L. (2005). 'Examining the relationship between learning organization characteristics and change adaptation, innovation, and organizational performance'. *Human Resource Development Quarterly*, 16(2), 185-211.

Kristensen, P.H. (2010). 'Transformative dynamics of innovation and industry: new roles for employees?' *Transfer European Review of Labour and Research*, 16(2), 171-183.

Lahiri, N. (2010). 'Geographic distribution of R&D activity: how does it affect innovation quality?' *Academy of Management Journal*, 53(5), 1194-1209.

Lawler III, E.E. (1986). *High-Involvement Management*. San Francisco: Jossey-Bass Publishers.

Liedtka, J.M. (1996). 'Collaborating across lines of business for competitive advantage'. *Academy of Management Executive*, 10(2), 20-34.

LO (The Danish Confederation of Trade Unions) (2007). Employee-driven-innovation – a trade union priority for growth and job creation in a globalised economy. (ISBN-online 978-87-7735-858-6). Available at: http://www.lo.dk/English%20version/Congress2007/~/media/LO/English/Congress2007/2007_1114_1227_Employee_driven_innovation.ashx (accessed 20 January 2011)

Love, J.H. and Roper, S. (1999). 'The determinants of innovation: R&D, technology transfer and networking effects'. *Review of Industrial Organization*, 15, 43-64.

Luecke, R. (2009). *The innovator's toolkit*. Boston, Massachusetts: Harvard Business Press.

Lukas, B.A. and Ferrell, O.C. (2000). 'The effect of market orientation on product innovation'. *Journal of the Academy of Marketing Science*, 28(2), 239-247.

McGill, M.E., Slocum, J.W. and Lei, D. (1992). 'Management practices in learning organizations'. *Organizational Dynamics*, 21, 5-17.

Marcotte, C. and Niosi, J. (2005). 'Small and medium-sized enterprises involved in technology transfer to China'. *International Small Business Journal*, 23(1), 27-47.

Møller, K. (2010). 'European innovation policy: a broad-based strategy?' *Transfer European Review of Labour and Research*, 16(2), 155-169.

Mowery, D.C. and Oxley, J.E. (1995). 'Inward technology transfer and competitiveness: the role of national innovation systems'. *Cambridge Journal of Economics*, 19, 67-93.

Nieto, M.J. and Santamaria, L. (2010). 'Technological collaboration: bridging the innovation gap between small and large firms'. *Journal of Small Business Management*, 48(1), 44-69.

Nijhof, A., Krabbendam, K. and Looise, J.C. (2002). 'Innovation through exemptions: building upon the existing creativity of employees'. *Technovation*, 22(11), 675-683.

Pedler, M., Burgoyne, J. and Boydell, T. (1991). *The Learning Company: a Strategy for Sustainable Development*. New York: McGraw-Hill.

Penrose, E.T. (1959). *The Theory of the Growth of the Firm*. New York:.Wiley

Porth, S.J., McCall, J. and Bausch, T.A. (1999). 'Spiritual themes of the "learning organization"'. *Journal of Organizational Change Management*, 12(3), 211-220.

Rocha, R.S. (2010). 'Shop stewards as coordinators of employee-driven innovation: implications for trade unions'. *Transfer European Review of Labour and Research*, 16(2), 185-196.

Schein, E. (1997). *Organizational Learning: What Is New?* Working Paper, MIT Sloan School of Management.

Senge, P.M. (1990). *The Fifth Discipline: the Art and Practice of the Learning Organization.* New York: Doubleday.
Shefer, D. and Frenkel, A. (2005). 'R&D, firm size and innovation: an empirical analysis'. *Technovation*, 25(1), 25–32.
Shuen, A. (1994). *Technology Sourcing and Learning Strategies in the Semiconductor Industry.* Unpublished PhD dissertation, University of California, Berkeley.
Spender, J.C. (1996). 'Making knowledge the basis of a dynamic theory of the firm'. *Strategic Management Journal*, 17, 45–62.
Sternberg, R. and Arndt, O. (2001). 'The firm or the region: what determines the innovation behavior of European firms?' *Economic Geography*, 77(4), 364–382.
Taherizadeh, A.H. (2010). *The Key Drivers of Innovation In Malaysia.* Unpublished master thesis, University of Malaya, Malaysia.
Teare, R. and Dealtry, R. (1998). 'Building and sustaining a learning organization'. *Learning Organization*, 5(1), 47–60.
Teece, D.J., Pisano, G. and Shuen, A. (1997). 'Dynamic capabilities and strategic management'. *Strategic Management Journal*, 18(7), 509–533.
Teglborg-Lefevre, A.C. (2010). 'Modes of approach to employee-driven innovation in France: an empirical study'. *Transfer European Review of Labour and Research*, 16(2), 211–226.
Telljohann, V. (2010). 'Employee-driven innovation in the context of Italian industrial relations: the case of a public hospital'. *Transfer European Review of Labour and Research*, 16(2), 227–241.
Terziovski, M. and Samson, D. (2007). 'Innovation capability and its impact on firm performance'. *Regional Frontiers of Entrepreneurship Research, AGSE Entrepreneurship and Innovation Research Exchange*, Australia, 362–374.
Therin, F. (2010). 'Learning for innovation in high-technology small firms'. *International Journal of Technology Management*, 50(1), 64–79.
Thompson, V. (1965). 'Bureaucracy and innovation'. *Administrative Science Quarterly*, 10, 1–20.
Tidd, J. and Bessant, J. (2009). *Managing Innovation: Integrating Technological, Market and Organizational Change.* Chichester: John Wiley and Sons.
Tidd, J., Bessant, J. and Pavitt, K. (1997). *Managing Innovation: Integrating Technological, Market and Organizational Change.* Chichester: John Wiley and Sons.
Tidd, J., Bessant, J. and Pavitt, K. (2001). *Managing Innovation: Integrating Technological, Market and Organizational Change.* 2nd ed. Chichester: John Wiley and Sons.
Tidd, J., Bessant, J. and Pavitt, K. (2005). *Managing Innovation: Integrating Technological, Market and Organizational Change.* 3rd ed. Chichester: John Wiley and Sons.
Yalabik, Z.Y., Chen, S.-J., Lawler, J. and Kim, K. (2008). 'High-performance work system and organizational turnover in east and southeast Asian countries'. *Industrial Relations*, 47, 145–152.
Yam, R.C.M., Guan, J.C., Pun, K.F. and Tang, E.P.Y. (2004). 'An audit of technological innovation capabilities in Chinese firms: some empirical findings'. *Research Policy*, 33, 1123–1140.
Wernerfelt, B. (1984). 'A resource-based view of the firm'. *Strategic Management Journal*, 5(2), 171–180.

11
Privileged Yet Restricted? Employee-Driven Innovation and Learning in Three R&D Communities

Tea Lempiälä and Sari Yli-Kauhaluoma

This chapter explores expectations, manifestations and constraints of employee-driven innovation (EDI) in communities that focus on technology development. Empirically, the study is based on an analysis of case studies on three different types of work communities in the domains of chemical technology, process technology and measurement technology. These kinds of communities can often be regarded as privileged regarding innovating because their members are expected to innovate as part of their daily work.

Currently, there is an ongoing demand for employees to be more innovative (Jamrog *et al.*, 2006) and organizations are required to bring inventions to the market at an ever-increasing pace (Miller *et al.*, 2005–2006). In this respect, R&D departments of established organizations as well as new high-tech ventures face particular pressures. Mostly, however, these pressures on employees and organizations seem to concentrate just on technological domains (Kessler and Chakrabarti, 1996; Markman *et al.*, 2005), which means that the R&D experts and firms are required to produce improvements or novelties as a result of technological innovation processes. This leads to the question of whether employee initiatives or workplace learning efforts outside technological development issues gain sufficient attention when activities are examined in the sphere of R&D in organizations. This question is highly relevant, since the extension of innovation initiatives and learning endeavours from technological domains into organizational development projects might have positive consequences not only regarding R&D outputs (Rousseau, 2011: 433) but also in aspects of employee motivation, engagement and even well-being (Garcia-Goñi *et al.*, 2007).

The purpose of this study is to explore innovating and workplace learning taking place in R&D contexts outside pure technological development work. We call this type of innovation and learning activities employee-driven innovation (EDI) (Høyrup, 2010). An important perspective in our study is that we regard R&D personnel not only as experts in technological domains but also as employees in their organizations who may have valuable insights and knowledge to develop and make innovations outside technological areas as well. To explore this suggestion, we conducted an empirical study in three different R&D communities in the areas of chemical technology, process technology and measurement technology. Our aim was threefold, in that we first tried to gain more understanding of whether R&D professionals in the communities that we studied face expectations to innovate outside technological domains. Second, we explored possible manifestations and the diversity of employee-driven innovation in R&D work communities. Finally, we analysed possible constraints of workplace learning and innovation initiatives in R&D communities. The results of our study suggest that R&D communities possess a lot of potential to innovate beyond technological domains as well, and organizations ought to pay more attention to the challenge of unleashing the full innovation potential of all their employees, including R&D personnel.

Employee-driven innovation in R&D contexts

While the concept of employee-driven innovation is still novel (Høyrup, 2010), the idea of including different professional groups or communities in the development of innovations in organizations has received a fair amount of attention. For example, Sundbo (1996) has indicated two systems of managing innovations in organizations: the expert system and the empowerment system. The former is often used in high-technology firms, whereas the latter is more typically in use in either service or low-tech firms. In the expert system organizations have a specific R&D department, whereas in the empowerment system the whole organization is encouraged to innovate. Sundbo (1996) argues for finding a balance between these systems, and states that much innovation potential is lost if the innovation capability and capacity of different communities within organizations are not mobilized in action. The inherent assumption here is that individuals and communities across organizations have valuable ideas and knowledge on how to develop their ways of working and in that way also produce innovations (Dougherty, 2004). Also, innovation capabilities in relation to the offering of the organization are believed to exist even outside R&D departments (e.g. du Chatenier *et al.*, 2009) and the utilization of these capabilities is generally encouraged. Interestingly, although this line of reasoning is fairly common (Atuahene-Gima, 1995; Sundbo, 1996), concerns over the utilization of the innovation capabilities of R&D professionals in domains other than technology are rare (Dougherty, 2004).

In this study we analyse innovation taking place within R&D communities. We use here the definition of work communities as spaces and places where people 'consider themselves to be engaged in the same sort of work'

(Van Maanen and Barley, 1984: 84). Our focus on R&D work communities is relevant here since, as Brown and Duguid (1991) point out, it is not only working but also learning and innovating that take place in such work communities. So far, studies examining the work of R&D communities have often taken various kinds of technology development perspectives (Hargadon, 1998; 2005; Hargadon and Sutton, 1997) and they have often centred on the questions of, for example, origins of creative output (e.g. Amabile and Gryskiewicz, 1988; Schoenmakers and Duysters, 2010), role of networks and collaboration behind technology development (Hansen, 1999; Powell *et al*., 1996) or improvement of performance or efficiency of technology development through either time management (Markman *et al*., 2005) or funding (Santamaría *et al*., 2010).

Thus, although there clearly exist various streams of research that shed light to the essence of the development of technological innovations, it seems that little attention has been paid to innovation activities conducted outside technological domains by R&D professionals. This raises the question of whether the expectations that R&D professionals face regarding innovation remain in some way narrowly determined and whether this poses barriers to the utilization of their creative capabilities. Limited expectations would be surprising, though, since managers are claimed to have high expectations of their company employees more generally to contribute to the development of innovations in organizations and workplaces (Reave, 2002). At the same time, it has to be noted that the linkage between learning and expectations has not been studied empirically in detail (Bosch-Sijtsema, 2007: 364). An additional aspect to note is that the expectations that R&D professionals set for themselves are often looked at only from the career development perspective (Cha *et al*., 2009) instead of examining the role of R&D experts in contributing to the development of their work communities or organizations (Dougherty, 2004).

Beside the expectations that employees in general (Jamrog *et al*., 2006) and R&D professionals in particular face from the management in workplaces regarding innovation, it is important to take a look at the actual manifestations of innovative activities within R&D communities. From the perspective of this study, we particularly aim to gain understanding of whether R&D professionals produce innovations outside technological domains, and, if so, what these innovations are like. To analyse and gain a better grasp of innovating, we follow in this study the practice-based perspective, which sees working, learning and innovating as 'interrelated and compatible and thus potentially complementary, not conflicting forces' (Brown and Duguid, 1991: 59). The practice-based perspective views 'learning as the bridge between working and innovating' (Brown and Duguid, 1991: 41) and puts emphasis on detailed investigation of work practices, since this offers us a way of gaining understanding of how learning and innovating take place in communities 'in the flow of experience' (Gherardi, 2000: 214). Bechky

(2006: 1760) crystallizes this idea by saying that 'in order to truly understand learning as a situated, social phenomenon, we must consider how it is embodied through work practice.' Similarly, Lave and Wenger (1991) emphasize the importance of gaining understanding of learning through analysing employees' participation in practice instead of their learning about practice. Learning, then, takes place in the course of an action, for instance, as people engage in reflection in their action (Schön, 1983). This kind of learning, then, may lead to innovations, while innovating in essence means changing practices in the course of action (Brown and Duguid, 1991).

So far, the existing research has identified at least two critical aspects related to practices that seem relevant when we aim to analyse innovating that reaches beyond the technological domains of R&D professionals. The one essential aspect concerns the understanding of R&D professionals of the content of their own work. According to Schön (1983), professional practice concerns not only solving problems, but also working out the whole context or specific situation around the problem to be solved. More precisely, the job of the professionals, besides problem-solving, includes defining 'the decisions to be made, the ends to be achieved, and the means which may be chosen' (Schön, 1983: 40). This kind of understanding of the spheres of professional practices in the field of R&D work then opens up possibilities for learning and innovating in workplaces that go well beyond developments in technological domains. The other relevant aspect related to R&D practices concerns how R&D work is organized in companies and firms. Dougherty (2004), for example, criticizes the traditional way of organizing R&D functions according to basic science or technologies, and recommends organizing R&D around practices instead. Dougherty (2004: 49) points out that organizing R&D around practices emphasizes 'what people should do and how they should make sense of their actions'. In essence, this kind of organization of R&D work emphasizes opportunities for R&D experts to learn and act beyond technological matters.

Finally, besides expectations to innovate and manifestations of employee-driven innovations, there may also exist critical constraints, either inside or outside work communities, that act as a hindrance to innovation activities (Amabile et al., 1996; Ekvall, 2000; Kanter, 1988). Manimala *et al.* (2006), for example, in their study of large public sector organizations have recognized several organizational constraints on innovating in R&D areas. Some of the constraints that they identified were, for instance, ambivalent support from immediate supervisor, inadequate rewards and recognition system, lack of recognition for innovations in non-core areas, lack of recognition for contributions by R&D support functions, and poor systems for promotion and management of ideas. Although the study of Manimala *et al.* (2006) focused on innovating within the field of R&D, their results point out that learning and innovating within and beyond R&D systems are intertwined and complementary systems within organizations. Also, Hargadon and Fanelli

(2002) point out that there may exist different types of knowledge within communities that may either enable or constrain employees' possibilities to learn and to innovate. All in all, then, many barriers must be overcome before innovating unfolds in organizations.

Research methods

Our study is based on the qualitative analysis and comparison of three independent case studies (Yin, 2003) on different types of R&D communities which we have called, according to their technological domains, Chemical community, Process community and Measurement community. The selection of these three cases is based on a meaningful combination of similarities and differences between the analysed communities. First, the three cases are similar in that they all focus on technology development on either a concept or an application level. Moreover, members of the studied communities are well educated in technological domains. The second reason behind the selection of our cases stems from the differences between the selected communities. Namely, the analysed communities varied in respect of the emergence of their existence, age and organizing principles. The combination of similarities and differences between the selected cases then allowed us to start making visible the spectrum of expectations, manifestations and constraints that R&D professionals may face regarding learning and innovating outside technological domains at their work. Table 11.1 presents the key facts on each R&D community in our study as well as the basic details of the research methods used.

In our three case studies, we used multiple methods in the collection of our empirical materials. The primary method in all our three cases was in-depth interviews. In this respect, a leading guideline in our studies was to search for and find the most 'knowledgeable informants' in each case (Eisenhardt and Graebner, 2007: 28). Therefore, we used the method of a 'snowball effect' (Balogun et al., 2005: 265), which meant that the interviewees themselves suggested other relevant potential interviewees. In the cases of Process and Measurement communities, some of the interviewees were also indicated by management. In addition to the interviews, we complemented our empirical material by written documents, visits to research and development facilities, and field observation. The use of field observation took place in the case of a Process community, while the analyses of Chemical and Measurement communities are mainly based on retrospective studies (Cox and Hassard, 2007) on concluded technology development projects.

Our analysis focused on innovation activities outside the technological domain and was guided by three questions. First, we aimed to gain understanding of whether the technological experts in our case communities faced expectations to innovate outside the technological domain, and, if so, where these possible expectations stemmed from. Our second question

Table 11.1 Three R&D communities

	Chemical community	Process community	Measurement community
Community type	Cross-organizational community	Established team within a large organization	Experimental structure (team) within a large organization
Technological context	Chemical technology	Process technology	Measurement technology
Community size	13 people	17 people	12 people
Background of community members	Six different organizations (from small start-up company to large established firms in various industries such as chemistry, food processing, pharmaceuticals, academia, and paper and pulp)	One large established organization	One large established organization
Education level of community members	Mixed (mainly higher education but also lower education and one master's student)	Higher education	Higher education
Existence of community	Emerging, evolving, temporary	Established, permanent	Newly founded, permanent
Age of community	Four years (collaboration ended)	10–15 years →	Two years →
Research methods	In-depth interviews, documents, visits to R&D facilities	Field observation, in-depth interviews, documents	In-depth interviews, documents, visits to facilities
Interviews	19 (from one to six hours long, recorded and transcribed)	20 (from 1.5 to 2.5 hours long, recorded and transcribed)	12 (from 1.5 to 2.5 hours long, recorded and transcribed)

explored how such innovation activities outside the technological domain would manifest themselves. Finally, we aimed to find out whether R&D professionals in the studied communities faced constraints regarding innovating beyond technological areas and, if so, what they were. To carry out the analysis, we studied the empirical data sets collected from each case individually (Staudenmayer et al., 2002: 587) to gain examples of expectations, manifestations and constraints of employee-driven innovation in the three cases. We then discussed and compared these examples to highlight the specificities of employee-driven innovation in each of our cases. Our process was highly iterative, since we constantly compared our preliminary findings with the existing literature.

Expectations, manifestations and constraints: results from three R&D communities

The results show that, although all three communities that we studied were active in the area of technological innovation, there were great differences in the community members' efforts concerning innovating outside their technological domains. Before presenting the key issues in each case, we have summarized in Table 11.2 our most important findings regarding expectations, manifestations and constraints on innovation beyond technological domains.

A cross-organizational community: chemical community

Chemical community was a cross-organizational community consisting of 13 chemical engineers, chemists or chemical technicians focusing on the development of one specific chemical innovation. Three of these chemistry professionals were scientist–entrepreneurs in a newly founded chemical high-tech venture, while ten of their collaborators were located in five large established organizations representing various industries: chemistry, food processing, paper and pulp, pharmaceutical manufacturing, and academia. The community originated because the scientist–entrepreneurs had invented a new type of recyclable chemical catalyst.[1] As an opportunity emerged to apply the catalyst in an industrially suitable chemical reaction, it led to the emergence of the Chemical community. The product development of the catalyst and process development of the chemical reaction then took place in the community among the scientist–entrepreneurs and their ten collaborators in five large established organizations. Moreover, the community tackled difficult scaling issues related to the transfer of product and process-related knowledge from small-scale laboratory environments into large-scale industrial production facilities. The accomplishment of the R&D development work took roughly four years, after which the community in the studied constellation came to an end.

Expectations concerning EDI: The members of the Chemical community represented different organizations in divergent market positions. These positions had an important influence on the varying expectations regarding learning and innovating that were put on different community members both by themselves and by their home-base organizations. On one hand, the scientist–entrepreneurs who were in the process of setting up a new business based on their technological invention were not only expected

Table 11.2 Results of cross-case comparison

	Chemical community	Process community	Measurement community
Expected innovation domain	Chemical technology	Process technology	Measurement technology
Expectations towards EDI within community	Mixed. Community members represented different organizations in divergent market positions. This led to varying expectations on the objectives of the community.	Indifferent. The community members were active innovators in the technology domain. In regards to the development of their own work procedures, they did not perceive their own role as important.	Frustrated. The members of the community were willing but not allowed to develop their ways of working. The leader wanted to protect the community and to keep the boundaries of the community closed.
Manifestations of EDI within community	Community members who were entrepreneurs used the community to develop their own business.	The leader of the group reflected on the ways of organizing and working of the group.	The community itself was an experiment in a new way of organizing and working.
Constraints on EDI within community	Requirements for economies of scale from background organizations. Established ways of working within each industry.	Only technological ideas included in the innovation domain. Organizational development excluded from the ideation activity. Limited possibility of implementing even technological ideas.	Hierarchical organization of the community. Restricted participation of the community members.

but forced to innovate outside the technological domain. For example, they needed to build a whole business organization and a production line for their invention from scratch. Moreover, the scientist–entrepreneurs had to search and find possibilities regarding where and how their invention, the catalyst, could be applied in industrial processes. They had no previous experience of such business activities due to their academic background. Therefore, they faced the challenge of learning how to operate in business and industrial settings. For instance, they had to be active in marketing and commercializing their chemical invention, which required them to search for potential industrial collaborators. Also, they had to learn to negotiate and secure funding for their start-up firm. The expectations regarding learning and innovating that the scientist–entrepreneurs confronted thus originated not only from themselves but also from their sponsors and potential business partners.

> Usually, everybody thinks that you just go ahead and sell [your product] to clients. But this is not so straightforward, because before your product can reach the client you ought to have some production processes [in operation]. And typically, clients have their own requirements regarding the product. Without understanding these requirements you cannot shape the product in its production process in a way that you could sell it to clients. (Scientist–entrepreneur, Chemical start-up firm)

As the comment of the scientist–entrepreneur above reveals, many of the requirements that they faced regarding learning and innovating were concentrated on the development of the technology itself. Nevertheless, to carry out these demands the scientist–entrepreneurs needed to learn how to build up and operate a business.

On the other hand, those members of the Chemical community who belonged to large, established organizations were in a very different position regarding expectations to learn and innovate outside technological domains. The expectations that they faced were mainly linked to either issues related to strategy of their home-based organizations or safety. These acted as a frame within which the community members were also allowed to operate within the Chemical community. Regarding strategies, these community members constantly reflected on their actions within the community in the light of how these would fit the strategic guidelines of their home-base organizations.

> It is part of the job that we constantly try to search for and test different kinds of ideas whether they will lead to something and check whether the ideas fit [the strategy of] the firm. (Senior expert, Chemical company)

Regarding safety issues, all people working in a technological field such as chemistry constantly face expectations to consider various risks involved in operations with chemical substances and processes. This means that safety issues constantly direct the ways in which things can and should be done in workplaces operating in the area of chemistry, and put pressure on continuous learning and innovating to maintain and improve safety in organizations. In this respect, continuous development of knowledge on how to operate various machines and pieces of equipment, as well as knowledge of how to deal with different kinds of chemical substances, plays an important role in work dealing with chemistry.

> We have to consider the safety requirements, practices and routines set by the machinery. ... We cannot improvise. Instead, we constantly have to consider safety matters and to make sure of the safety. (Senior expert, production unit of a paper and pulp company)

Manifestations of EDI: There were several manifestations of learning activities outside technological domains among the scientist–entrepreneurs. The most important example was the search for prospective partners or customers for the technological applications of their invention. The scientist–entrepreneurs created innovative ways to spread information and to raise interest in their technology. For instance, they delivered samples of their invention, the chemical catalyst, to potentially interesting partners and conducted tests of the catalyst in different kinds of conceivable chemical applications. They also created small-scale joint research projects with prospective partners. Basically, all of these were innovative, tailor-made marketing solutions to finding a business potential for their invention. It is important to note here that marketing was an area that had previously been essentially unknown to the scientists. Marketing, in the form of increasing knowledge of the invention among potential collaborators and clients and thereby creating business opportunities, was something that the scientist–entrepreneurs had to learn to do. Another relevant example of their learning activities outside the technological domain was the construction of a large-scale production line. Whereas building machines for their production line itself fell within the domain of their expertise, they also learned to negotiate new kinds of deals with larger, established organizations that allowed them to utilize existing machinery for the start-up business. The scientist–entrepreneurs thus created novel organizational solutions, by using and developing their relations in an innovative way for the development of their manufacturing processes. The resulting production line then formed the core of their start-up firm, which the scientist–entrepreneurs saw as providing them with both confidence in being able to operate in an industrial setting and credibility as a business partner.

Constraints of EDI: Two important constraints that seemed to hinder at least some of the learning potential in the Chemical community were related to safety issues and existing machinery. On the one hand, the strict safety norms and rules direct not only the technology development but also the way things are done and can be done in the industrial context. On the other hand, the existing facilities and machinery needed in chemistry are typically expensive and involved with rigid organizational and operational processes. Due to these factors, changing a chemical production process is considered as a risk, both economically and for safety reasons.

> 'There must be strong reasons to make changes [in a production process]. Basically, [the only reason is] that your system does not work in practice.... Everyone knows that saving like a million by applying a new system may cost you ten millions in case it does not work in practice and therefore your production is suspended. (Scientist–entrepreneur, chemical start-up firm)

In addition to the industry-specific safety and machinery issues, those members of the Chemical community who were located in large organizations seemed to face one particular constraint on learning and innovating at work. This was the requirement set by the management to meet the criteria for the economies of scale. In essence, this means that, if management mainly recognizes and encourages innovations that lead to radical technological changes or to large sale volumes, we can raise the question of whether people in these kinds of organizations are recognized or even allowed to make small-scale improvements outside technological areas that still might be somehow meaningful and important from the perspective of learning at work. As already noted earlier, learning and innovating are endeavours full of uncertainties (Hardadon and Fanelli, 2002), and therefore the end results of learning may not always be predictable. Hence, the emphasis of management on measurable and predictable results may sometimes cause frustration, as, for example, was the case expressed by some members of large organizations in the Chemical community.

> Even today, you can find several [examples on occasions] where people think, that things will proceed in certain ways. These expectations may be based on calculations, even exact ones. But ... it is very difficult to see all the consequences of particular investments. ... Sometimes there might even be some negative consequences but we can find out about these only at a later stage. ... Hardly anyone gives funding for those kinds of projects where the end result is most likely negative, but where the project itself would be an important learning experience for the personnel. (Senior expert in an established firm, Chemical community)

An established team within a large organization: process community

Process community is a technology team of 17 professionals in a global technology company. The company provides technology and services within a traditional process industry dealing with metals processing. The Process community has a long history and an important position in the company, since it develops one of the core technologies of the company. In addition to technology development, the community consults other organization members in issues related to this technology. The members of the Process community are highly educated in various technological domains. The professional ages of the team members vary from few years to three decades.

Expectations concerning EDI: The work of the Process community is strongly concentrated on the core technology of the firm in the metallurgy business. For example, the community members solve problems concerning the applications of the company's core technologies in customer facilities or create new solutions to better fit customer needs. The work of the technology experts also includes more routine-like tasks, such as project documentation and internal consultation, but the creation of development ideas in the technological domain is a central part of their work. The innovation expectations that the community members face emerge through either customer projects or their own internal technology development endeavours. Although the technology experts work closely in customer projects, the expectations of their contribution to these projects are largely technological. They take part in sales efforts as well as implementation projects at customer facilities, but they are expected to act as technological experts rather than providing ideas of, for example, new business models or making explicit sales efforts. The customer projects are managed by the project organization and the technology experts are individually assigned to the projects according to need, although at times the customer contacts the process community directly. In the latter case the technology expert contacts someone from either the sales or project organization in order to get a management or business perspective on the project. Internal technology development efforts, however, are managed autonomously by the Process community. The process community members are thus in frequent contact with people from other parts of the organization, but their role is always related to providing technological expertise and they are rarely asked to contribute to other domains. In summary, the community members are highly motivated and active in technology development, which is also expected of them. However, innovating outside the technological domain is not expected of them, and they do not show particular enthusiasm for it either.

Manifestations of EDI: The only member of the Process community who showed interest in innovating outside the technology domain was the team leader. More particularly, the team leader was interested and active in developing the working procedures of the team, such as ideation techniques. Although this type of activity is often expected from team leaders in this studied organization, it was not set as a particular requirement for the leader of a technological team. Other community members stated that they were proactive towards the development of their own work and claimed to search for ways to do things better. However, most of these kinds of development efforts were related to creating new technological solutions regarding the end product as opposed to developing internal work procedures.

> Innovating is extremely important at my own work. ... I think I have about five or six [patents] and a few invention notices in process. There was some kind of an idea management or initiative system [in our firm] but I haven't even checked the status of the system for ages. [A laugh] (Senior Expert, Process Inc.)

An indirect manifestation of the innovating and learning endeavours of Process community members in non-technological areas, however, was the heavy engagement of technology experts in communication with customers. In essence, the technology experts had most of their ideas when interacting with customers; listening and learning about their problems in technology and business and familiarizing themselves with the customer facilities.

> Well, if I think about my colleagues in our team, basically all of us have a possibility to gain access to the information [concerning the problems that our customers face]. We all deal with the sales function and interact so much with our customers that we constantly [learn about] challenges that require further development. (Senior expert, Process Inc.)

Nevertheless, the problems that the Process community members identified and ideas that they created for the customers were always technological, and, therefore, possibilities to innovate and learn in relation to customers' business solutions or other non-technological issues were neither directly offered to nor sought by the R&D experts.

Constraints of EDI: The Process community members faced significant time pressure due to the high number of customer projects. They had to struggle to be able to invest time even in technological development efforts. Nevertheless, as the technological development accomplishments were clearly their main priority at work, they managed to organize time for it. Development efforts outside the technological domain were, however, not a priority for them in any dimension, and therefore they did not allocate time for this type of activity.

> There is no time to develop tools or methods. It is a challenge. (Senior expert, Process Inc.)

Besides time pressure, an important constraint on innovation efforts outside technological domains stemmed from the expectations set for the community by the management and by the members themselves. In a word, innovation efforts outside technological domains were simply not expected or encouraged by the management. Even more, the community members themselves did not perceive the potential value of their contribution in non-technological domains and did not even consider these kinds of endeavours as part of innovation activity. Although the members of the Process community strongly identified themselves with terms such as innovation and innovativeness, these terms had solely technical connotations for them, as can be seen in the following quote.

> Being able to sell ideas ... does not actually mean innovating; it is just hard work or requires pushing ... Innovation means something that you have first made a proposal for. In the next phase, it will be decided whether [the invention] will be patented or what the company will make with it. If they decide that the invention will be patented, the patent agent will make a contact with you. (Senior Expert, Process Inc.)

In essence, the quotation of the senior expert above means that innovating outside technological domains, such as creation of business models or development of work practices and procedures, was completely excluded from their perspective. These perceptions were strengthened by the strong influence of intellectual property rights in the organization; guiding all development activities and dominating the discourse on learning and innovating within the firm. Moreover, the channels through which non-technological employee initiatives were handled in Process Inc. were considered by the R&D experts in the firm to be unfamiliar, even irrelevant, compared with the intellectual property rights processes. This is most likely due to the fact that the employee initiative systems and related campaigns were usually targeted to other parts of the organization than the R&D function.

An experimental team: measurement community

Measurement community is a concept development team in a relatively large globally operating company that produces measurement-related products for various industries such as environment and energy. The studied team is located within one of the three business divisions and represents a novel structure within the whole organization. The purpose of the team is to develop innovative concepts which they then offer for further development and commercialization to one of the business divisions. The concepts can be related to either measurement products, services or business models. The team consists of 12 technology professionals in different areas, such as ICT technology and electrical engineering. The team is managed by a team leader, who originally assembled the team, and a project manager, who is in charge of the concept development projects.

Expectations concerning EDI: The original vision for the Measurement community was to create a group that could produce new kinds of business concepts – albeit from technological starting points – for the company. There were high expectations directed towards the team: the management expected innovative outcomes and the team members anticipated challenging and diverse work tasks that would allow them to utilize their full creative potential. For the team members, this also meant stepping out of their immediate technological expertise area and contributing to the concept development entity. However, the team leader and project manager had differing views of the role of the team members and tended to allocate them tasks related to their immediate expertise area, while keeping the control of the concept entity as well as customer contacts to themselves. This created frustration among the team members and affected their enthusiasm and motivation to conduct their work task and learning at work. This is well illustrated in the following quote from a team member's interview:

> It [that is the big picture of the concept] was already there. I think that this was probably a big de-motivating factor [for innovating] right from the start.... It would be great to do real conceptualization work... that we would have a problem that would be open [for development] to all of us. (Team member, Measurement Inc.)

In brief, the expectations of team members' innovation activities outside their immediate technological domains were thus low on the part of team management but high on the part of the team members themselves.

Manifestations of EDI: The concept development projects that were handed over to the team members were usually carefully prepared by the team leader and project manager beforehand. The normal way to proceed with a concept was that these two managers first prepared three alternative concept ideas and then presented these to the management board. Additionally, these two managers were involved in the selection of one of these prepared concepts to be developed by the team. Usually, the team members then did not have a say in choosing their development tasks regarding the concept or creating the initial concept idea. Only once, in the course of the first development project, was the concept idea worked on and transformed as a joint effort of all participants, and all the team members were able to participate, for example, in visits to customer sites to identify requirements for innovating regarding the project. This was important to the team members, since the personal contact with the customer facilities allowed them to both gain an understanding of and have an influence on the business model development of the concept. The team members found this highly motivating, and the possibility of contributing to the concept entity made them feel ownership of the project and think about development efforts beyond their immediate technological expertise.

Constraints of EDI: The most important constraint on innovating outside technological domains in the Measurement community was the control exerted by the management duo on the development and choice of the concepts. The team members were not able to utilize their full creative potential beyond their particular technological domains, largely because the management duo perceived them as technical experts but not as having the potential or interest to be engaged in ideating the business side of the concept. In fact, when specifically asked why the technology experts were not included in the business side development of the concept, the management duo replied that they did not believe that the R&D experts would be motivated to take part in such kinds of activities. The management was under the impression that the R&D experts would like to focus only on technological matters and that they would not like to be bothered by other issues. In contrast, the team members felt that their potential was not fully utilized if they were kept out of the wider concept development activity. This discrepancy was a source of frustration to many team members, as the situation was in conflict with the way they perceived their own role in the team.

Discussion

The aim of our study was to explore expectations, manifestations and constraints on employee-driven innovation in communities that focus on technology development. Essentially, that is to say that we aimed to study conditions and accomplishments of R&D communities' innovation activities reaching beyond their technological domains. It is important to note here that we do recognize that the boundary between technological and non-technological innovating is not clear-cut. Therefore, it is important to clarify that our intention was to study how far R&D professionals can and do reach beyond their own technological expertise at their work and participate in the development of their jobs, workplaces and organizations.

The results of our three case studies indicate that the expectations that R&D experts face regarding innovating and learning outside their technological domains are rather ambiguous. Those R&D professionals who set up new businesses are not only expected but obviously forced to expand their expertise beyond technological knowledge and know-how in order to learn to make and develop a new business. On the other hand, the expectations regarding non-technological innovation of those R&D professionals who are based in established, large organizations seem to be much more restricted – either by their own perceptions or by management. Nevertheless, we did also find some cues referring to innovation activities beyond technological domains in R&D communities in large established organizations. For example, the head of the technology group in the Process community was active in developing the working procedures of the team, such as ideation techniques. Additionally, all technology experts in the Process community actively engaged in interaction with customers, which at least offered them opportunities to reach beyond technological issues. A similar kind of interaction taking place at least occasionally in the Measurement community stimulated technology experts to contribute to innovating beyond their immediate technological expertise. Despite these activities, though, we recognized many constraints on employee-driven innovation in technology communities, such as rigid operational procedures, emphasis on technological innovations in general and those resulting in economies of scale in particular, lack of time, and management control.

Our results, then, suggest that technology experts – though privileged in the technological innovation domain – are at the same time restricted in their innovation activities reaching beyond technological domains. This indicates that employees in R&D functions may have creative potential that is currently not fully utilized by their workplaces and organizations. This raises several questions concerning both the organization of technological work (Dougherty, 2004) and the role of R&D professionals in organizations. Thus, the discussion of how R&D personnel can be both involved and recognized in innovating beyond technological domains seems a relevant topic to study in more detail in the sphere of discussion on employee-driven innovation.

Note

1. Catalysts are materials that change the rate of attainment of chemical equilibrium without themselves being consumed in the process. They are one of the core technologies for the manufacture of chemicals and materials, and widely used in several industries from food processing to pharmaceuticals (Wittcoff and Reuben, 1996; Adams, 1999).

References

Adams, C. (1999). 'Catalysing business'. *Chemistry and Industry*, 19: 740–742.
Amabile, T.M., Conti, R., Coon, H., Lazenby, J. and Herron, M. (1996). 'Assessing the work environment for creativity'. *The Academy of Management Journal*, 39, 1154–1184.
Amabile, T.M. and Gryskiewicz, S.S. (1988). 'Creative human resources in the R&D laboratory: how environment and personality impact innovation.' In Kuhn, R.L. (Ed.) *Handbook for Creative and Innovative Managers*. New York: McGraw-Hill.
Atuahene-Gima, K. (1995). 'An exploratory analysis of the impact of market orientation on new product performance: a contingency approach'. *Journal of Product Innovation Management*, 12 September, 275–293.
Balogun, J., Gleadle, P., Hailey, V.H. and Willmott, H. (2005). 'Managing change across boundaries: boundary-shaking practices'. *British Journal of Management*, 16(4), 261–278.
Bechky, B.A. (2006). 'Talking about machines, thick description, and knowledge work'. *Organization Studies*, 27(12), 1757–1768.
Bosch-Sijtsema, P. (2007). 'The impact of individual expectations and expectation conflicts on virtual teams'. *Group & Organization Management*, 32(3), 358–388.
Brown, J.S. and Duguid, P. (1991). 'Organizational learning and communities-of-practice: toward a unified view of working, learning, and innovation'. *Organization Science*, 2(1), 40–57.

Cha, J., Kim, Y. and Kim, T.-Y. (2009). 'Person-career fit and employee outcomes among research and development professionals', *Human Relations*, 62(12), 1857–1886.
Cox, J.W. and Hassard, J. (2007). 'Ties to the past in organization research: a comparative analysis of retrospective methods'. *Organization*, 14(4), 475–497.
Dougherty, D. (2004). 'Organizing practices in services: capturing practice-based knowledge for innovation'. *Strategic Organization*, 2(1), 35–64.
du Chatenier, E., Verstegen, J.A.A.M., Biemans, H.J.A., Mulder, M. and Omta, O.S.W.F. (2009). 'The challenges of collaborative knowledge creation in open innovation teams'. *Human Resource Development Review*, 8(3), 350–381.
Eisenhardt, K.M. and Graebner, M.E. (2007). 'Theory building from cases: opportunities and challenges'. *Academy of Management Journal*, 50(1), 25–32.
Ekvall, G. (2000). 'Management and organizational philosophies and practices as stimulants or blocks to creative behavior: a study of engineers'. *Creativity and Innovation Management*, 9(2), 94–99.
Garcia-Goñi, M., Maroto, A. and Rubalcaba, L. (2007). 'Innovation and motivation in public health professionals'. *Health Policy*, 84(2–3), 344–358.
Gherardi, S. (2000). 'Practice-based theorizing on learning and knowing in organizations'. *Organization*, 7(2), 211–223.
Hansen, M.T. (1999). 'The search-transfer problem: the role of weak ties in sharing knowledge across organization subunits'. *Administrative Science Quarterly*, 44(1), 82–111.
Hargadon, A.B. (1998). 'Firms as knowledge brokers'. *California Management Review*, 40(3), 209–227.
Hargadon, A. (2005). 'Technology brokering and innovation: linking strategy, practice, and people'. *Strategy & Leadership*, 33(1), 32–36.
Hargadon, A. and Fanelli, A. (2002). 'Action and possibility: reconciling dual perspectives of knowledge in organizations'. *Organization Science*, 13(3), 290–302.
Hargadon, A. and Sutton, R.I. (1997). 'Technology brokering and innovation in a product development firm'. *Administrative Science Quarterly*, 42(4), 716–749.
Høyrup, S. (2010). 'Employee driven innovation and workplace learning: introduction of basic concepts, approaches and themes'. *Transfer*, 16(2), 143–154.
Jamrog, J., Vickers, M. and Bear, D. (2006). 'Building and sustaining a culture that supports innovation'. *Human Resource Planning*, 29(3), 9–19.
Kanter, R.M. (1988). 'When a thousand flowers bloom: structural, collective and social conditions for innovation in organization'. *Research in Organizational Behavior*, 10, 169–211.
Kessler, E.H. and Chakrabarti, A.K. (1996). 'Innovation speed: a conceptual model of context, antecedents, and outcomes'. *Academy of Management Review*, 21(4), 1143–1191.
Lave, J. and Wenger, E. (1991). *Situated Learning. Legitimate Peripheral Participation*. New York: Cambridge University Press.
Manimala, M.J., Jose, P.D. and Thomas, K.R. (2006). 'Organizational constraints on innovation and intrapreneurship: insights from public sector'. *Vikalpa: The Journal for Decision Makers*, 31(1), 49–60.
Markman, G.D., Gianiodis, P.T., Phan, P.H. and Balkin, D.B. (2005). 'Innovation speed. Transferring university technology to market'. *Research Policy*, 34(7), 1058–1075.
Miller, L., Miller, R. and Dismukes, J. (2005/2006). 'The critical role of information and information technology in future accelerated radical innovation'. *Information Knowledge Systems Management*, 5(2), 63–99.

Powell, W.W., Koput, K. and Smith-Doerr, L. (1996). 'Interorganizational collaboration and the locus of innovation: networks of learning in biotechnology'. *Administrative Science Quarterly*, 41(1), 116–145.

Reave, L. (2002). 'Promoting innovation in the workplace: the internal proposal'. *Business Communication Quarterly*, 65(4), 8–21.

Rousseau, D.M. (2011). 'Reinforcing the micro/macro bridge: organizational thinking and pluralistic vehicles'. *Journal of Management*, 37(2), 429–442.

Santamaría, L., Barge-Gil, A. and Modrego, A. (2010). 'Public selection and financing of R&D cooperative projects: credit versus subsidy funding'. *Research Policy*, 39(4), 549–563.

Schoenmakers, W. and Duysters, G. (2010). 'The technological origins of radical innovations'. *Research Policy*, 39(8), 1051–1059.

Schön, D.A. (1983). *The Reflective Practitioner. How Professionals Think in Action*. New York: Basic Books Inc.

Staudenmayer, N., Tyre, M. and Perlow, L. (2002). 'Time to change: temporal shifts as enablers of organizational change'. *Organization Science*, 13(5), 583–597.

Sundbo, J. (1996). 'The balancing of empowerment – a strategic resource based model of organizing innovation activities in services and low-tech firms'. *Technovation*, 16(8), 397–409.

Van Maanen, J. and Barley, S.R. (1984). 'Occupational communities: culture and control in organizations'. In Staw, M.B. and Cummings, L.L. (Eds.) *Research in Organizational Behavior* (Vol. 6) (pp. 287–365). Greenwich: JAI Press.

Wittcoff, H.A. and Reuben, B.G. (1996). *Industrial Organic Chemicals*. New York: Wiley.

Yin, R.K. (2003). *Case Study Research. Design and Methods*. Thousand Oaks, California: Sage Publications.

12
Employee-Driven Innovation and Industrial Relations

Stan De Spiegelaere and Guy Van Gyes

Both industrial relations and innovation are well-established subjects in the current scientific literature. Although research has frequently related the two concepts, it has rarely focused on or considered employee behaviour. This chapter reviews the literature linking Employee-Driven Innovation with two key concepts of the industrial relations field: employee participation through workplace representation and collective bargaining outcomes such as wage and employment regulation. This chapter concludes that direct participation is positive for EDI; indirect participation stimulates direct participation and can positively influence EDI when embedded in optimal company industrial relations. Further, the literature review uncovers a general lack of empirical research on the effects of labour regulation and wages on EDI and related employee behaviour.

Introduction

Innovation is currently seen as the key to sustained economic performance of European nations and firms. Along with traditional innovations rooted in R&D and entrepreneurship, the innovative potential of employees is currently being valued more and more as an important source of innovation. Literature on how to stimulate this 'Employee-Driven Innovation' (EDI) or innovative work behaviour of employees is booming. The context in which the employee works is an essential factor in explaining employee behaviour, and a crucial aspect of this context is the 'employment relationship' between the employer and the employee, which is formed through and by the industrial relations (IR) in the company. IR determines the conditions in which an employee is engaged and affects the climate at the workplace. Therefore, IR can be rightly considered to affect employee behaviour in the field of innovation. Nonetheless, research only rarely focuses on different aspects of industrial relations and their link with employee behaviour (Van Gyes, 2003). This chapter reviews this limited stream of literature on the subject of industrial relations and innovation in search of indications about how IR affects EDI.

Industrial relations

Industrial relations is the area of study that focuses on 'the governance of the employment relation in its totality, along with its economic, political and social implications' (Sisson, 2008: 45). The 'employment relationship' further is defined as the 'legal creation in which one person (the employee) agrees for a sum of money specified over some time period to provide labour to another person (the employer) and follow the employer's orders and rules regarding the performance of work, at least within limits' (Simon, 1951, in Kaufman, 2004: 51). The employment relationship can hence be divided into two separate, but linked, dimensions. On the one hand, and at its most basic level, the employment relationship is a matter of economics. Individuals offer their skills and abilities to an employer for a price. Economic considerations, such as wages and other benefits, are major factors in individual and firm decisions to establish the employment relationship (Block *et al.*, 2004). On the other hand, the relationship also has a social dimension, which is about the subordination of the employee and the authority of the employer. Central here is how this hierarchical relation is structured; how control is exercised, managed and organized.

The governance of this employment relationship – industrial relations – in the Western market economies, and in the European social model especially, has gone through a process of democratic institutionalization. Statutory frameworks, trade union recognition and supra-company regulation are key features of the industrial relation systems that developed throughout Europe (Hyman, 2000).

Two institutional features have been central in the development of these systems. The first concerns the *external, economic, contractual* aspects of the job, such as employment status, wage and working time. The representatives of employees (unions) and employers *negotiate* these aspects of the employment relation in *collective bargaining*. In many European countries relatively centralized and coordinated forms of collective bargaining have been established. The second aspect of the employment relation concerns the *job itself* and how it's supposed to be performed. As already stated, the employee agrees to respect his *subordinate position* and act according to the directions of the employer, while the employer agrees to *inform and consult* the employees and their representatives in relation to the organizational management on a regular basis. This right of information and consultation is sometimes also referred to as 'indirect participation in the workplace'. A core feature of this right in European IR systems is the integration of labour into managerial decisions through statutorily recognized structures of employee representation, such as a works council or union shop steward. These bodies have to guarantee the right to information and consultation (cf., at EU level, the directive of 2002).

The IR system, its structure and practice are directly targeted at influencing the crucial employment relation in which the employee is tied to the employer, and vice versa. Participation in managerial decisions by employee representation and collective wage bargaining are key elements of this system, certainly in the European tradition. Although there is a relative absence of studies relating IR to EDI, with the article of Telljohann (2010) as a notable exception, this contribution gives an overview of the existing literature, pinpoints important blanks and concludes with some research opportunities. First the literature which links employee representation and participation with employee behaviour will be examined; next the literature on the main outcomes of collective bargaining – wages and job security – and their effect on EDI.

Innovation, innovative work behaviour and EDI

The terms 'innovation' and 'innovative employee behaviour' are defined according to West and Farr (1990) as 'the intentional introduction and application, within a role, group or organization of ideas, processes, products or procedures, new to the relevant unit of adoption, designed to significantly benefit the individual, the group, organization or wider society'. Innovative work behaviour (IWB) is the behaviour of employees which involves not only the creation of an idea, the discovery of something, but also the introduction and application of that idea with the intention to provide a benefit (de Jong and Den Hartog, 2010: 6). The concept of EDI goes further and refers to the idea that employees are crucial actors for innovation in organizations (Høyrup, 2010). They are central to the implementation phase but also, and more importantly, to the pre-design and design phases of the innovation. Employees frequently face concrete problems that can be solved through workplace innovation and they are in a unique position to assess whether proposed solutions and innovations are practically applicable.

Employee representation, participation and innovation

Workplace social dialogue is an IR process whereby recognized employee representatives are involved in decisions concerning the employment relationship at the workplace (Van Gyes, 2010). Such involvement may be limited to just being informed by management, or may extend to consultation, negotiation or joint participation in decision-making. The basic structure is through union representation / shop steward or a more general works council type. Works councils are legally established representations, elected or appointed by all employees at an establishment, irrespective of their membership in a trade union.

In the literature we find a range of studies linking forms of employee representation with innovation performance of companies. However, the link with IWB or EDI is only rarely made (see Table 12.1). The link between forms of direct, task-based employee participation and EDI, on the contrary, is strongly established. As a kind of third relationship, we find a large amount of literature linking these forms of direct, task-based participation and employee representation. In the following section, we discuss these (non-)established links in the literature in more detail.

Employee representation

In Tables 12.1 and 12.2 a list of research literature on the relation between employee representation and innovation is given. As already stated, and as an important observation, the innovation-related literature on employee representation mainly makes links with general innovation input or output indicators, without referring to elements of EDI.

When reviewing the literature, furthermore, we notice the following. First, the literature can largely be split into two categories, one that focuses on the effects of unions on innovation and one that looks into the relation between works councils and innovation. Second, the research results concerning the effects of works councils are predominantly based on German observations, with notable exceptions from the UK and the Netherlands. Germany is, of course, the birthplace and host country of a well-established type of works council. The research on the effect of unions is geographically more diverse. Third, the research frequently uses dummy variables for works councils and unions. Research that also measures the activity of works councils, the type of industrial relations climate or attitudes of works council members and management is rare. Fourth, there's no research that measures the effect of works councils or unions on employee behaviour.

When we compare the *outcomes*, the inconsistency of the results is striking. Even when we distinguish between union and general employee representation research or split up the research per country, no consistent results emerge. Regarding the *works councils*, both positive and negative relations are found. Two interesting pieces of research can give us a clue about the reasons behind these inconsistent results. Notably, the research of Dilger (2002) 'deepened' the works council variable and observed that active works councils were indeed positively related to innovation. A more recent study by Jirjahn and Smith (2006) distinguished between four situations, depending on the presence/absence of a works council and the attitude of the management towards workers' involvement in companies. They also distinguished between different types of product innovation. The combination of a works council and positive management attitudes had the largest effect on the introduction of products with 'improved quality or additional functions'. According to the researchers, this can primarily be explained by changed employee behaviour. It is not only the presence of works councils that seems to matter; the type of works council, its activity and the industrial relations climate in the company also determine the efficacy of the works council in

Table 12.1 Works councils and innovation

Authors	Sample	Indirect participation	Innovation measurement	Main findings	Country
Addison and Schnabel (1996)	1,025 firms	Dummy and workplace representation index	Introduction of new products or processes (survey)	Insignificant for the dummy, positive for the workplace index for product innovation. Insignificant for process innovation	Germany
Addison et al. (2001)	900 firms	Dummy	New processes and products	Insignificant relation	Germany
Addison et al. (1993)	50 est.	Dummy	Profitability, value added and investment	Insignificant relation except with investment in physical capital (negative)	Germany
Dilger (2002)	1,716 firms	Dummy	Product innovations	Positive but insignificant relation Positive association when works councils are strongly involved in the decision-making	Germany
FitzRoy and Kraft (1990)	57 metal firms	Union density and WC dummy	Sales of new products of the last five years	Negative and significant	Germany

Hübler (2003)	78 firms	Dummy	Innovations	Positive and significant	Germany
Schnabel and Wagner (1992)		Dummy	Product innovation	Positive but insignificant	Germany
Jirjahn and Smith (2006)	709 firms	Work council and attitude of management	Various types of product innovations	Positive attitudes and council positively related to improved quality and/or additional features	Germany
				Positive attitudes positively related to completely new products	
				Negative attitudes and council negative relation with improved quality but positive with other innovations	
Wigboldus et al. (2008)	Three case studies	Dummy	Performance, profitability, innovation	Work councils can be a strategic partner and result in enhanced profitability and performance	Netherlands

promoting innovation. It can therefore be concluded that the presence of a works council alone doesn't automatically lead to higher levels of EDI, but an active works council embedded in good company industrial relations does increase innovative behaviour of employees; just as previous research already found that a cooperative industrial relations climate (Blyton et al., 1987) positively influences organizational commitment and therefore organizational change (Iverson, 1996).

Concerning the effect of *unions* on innovation, most research focuses on the negative, indirect effects of unions through the increased price of labour, decreased profitability and hold-up problems between managers and unions, which would undermine the investment motive and entrepreneurs' capability to innovate (Menezes-Filho and Van Reenen, 2003). Some also point to the possible positive 'voice' effect of unions (Freeman and Medoff, 1984), but this is rarely taken into account. The research results of the literature presented in Table 12.2 are not conclusive. There is research indicating positive, negative and no relations. When positive effects are reported, the research mostly observes that unions reduce employees' anxiety and resistance towards innovations. An active role for the union as a promoter of workplace innovation and employee experimenting is rarely researched, although research shows that union cooperation makes, for example, the introduction of high-involvement human resources management more successful (Cooke, 1994; Gollan and Davis, 2001; Roche and Geary, 2002; Therrien and Leonard, 2003). These studies show that the involvement of unions in the decision-making process increases not only acceptance of the changes but also the efficacy of the innovation, as employee knowledge is mobilized through the unions.

Direct participation and innovation

In contrast to employee representation in managerial decisions, direct participation in task-based decisions and innovative employee behaviour have been researched more thoroughly. The literature is rather straightforward. Various studies with different methodologies in different countries indicate positive effects of direct participation on innovation, innovative work behaviour and different concepts that are close to the EDI concept. We refer to the literature linking participation with organizational innovation (Dhondt and Vaas, 1996; Guthrie et al., 2002; Kivimaki et al., 2000; Laursen, 2000; Laursen and Foss, 2003; Lay, 1997), innovative behaviour of employees (Chen and Aryee, 2007), organizational citizenship behaviours (Bogler and Somech, 2005; Cappelli and Rogovsky, 1998; VanYperen et al., 1999), knowledge-sharing (Han et al., 2010), and even EDI (Telljohann, 2010). We can therefore conclude that direct participation is a successful way of promoting EDI.

Forms of direct participation are a central component of the 'innovative' organization. Direct participation intensifies and enlarges knowledge flows

Table 12.2 Unions and innovation

Authors	Sample	Indirect participation	Innovation measurement	Main findings	Country
Rogers (1999)	AWIRS	Union presence Union density	Three categories of innovative companies	No relations	Australia
FitzRoy and Kraft (1990)	57 metal working firms	Union density and work council activity	Sales of new products of the last five years	Negative and significant	Germany
Blundell et al. (1999)	Firm-level panel data	Industry union density	Innovation (survey data)	Negative effect	UK
Geroski (1990)	73 industries	Number of workers covered by a collective agreement	Number of technically and commercially successful innovations	Negative but insignificant	UK
Machin and Wadhwani (1991)		Union recognition Presence of JCC	Investments	Positive and significant Positive and significant	UK
Michie and Sheehan (2003)		Union density, Dummy variable (50%)	Product and process innovation – survey response	Positive and significant	UK
Acs and Audretsch (1987)	247 industries	Union density	Number of innovations	Negative effect	US
Hirsch and Link (1987)	315 firms	Union presence; Dummy (50%)	Response data	Negative effect	US

Source: AWIRS: Australian Workplace Industrial Relations Survey; JCC: Joint Consultative Committee.

because of better vertical decentralization, horizontal coordination and organizational commitment. Employees have to be given the opportunity to put their knowledge to use in the workplace. Involving employees in decisions that affect day-to-day tasks helps to create a culture of autonomy and responsibility. Employers and managers need to be receptive to feedback and suggestions. In this way direct participation creates an organization of high involvement that spurs innovative work behaviour of employees.

Direct task-based participation and employee representation

Other research about the interplay of direct and indirect participation has found that they are related. Companies with indirect participation schemes generally have more forms of direct participation (Addison and Belfield, 2003; Sisson, 1993). OECD studies confirmed this and saw that high-involvement human resources strategies that encourage direct participation are more likely in workplaces covered by collective agreements and are related to industrial relations systems that favour cooperation between employers and employees (OECD, 1999). Also Black and Lynch (2004) found that employee voice and involvement produced larger effects in unionized companies than in non-unionized companies.

Employee representation and EDI: conclusion and discussion

To conclude this first section: employee representation in managerial decisions alone does not change the innovativeness of companies and employees, but when embedded in cooperative industrial relations it can produce positive effects. Next, direct participation is positively related to innovative employee behaviour, and a quality employee representation reinforces direct participation. In short, sufficient scientific proof has been established to show that both indirect representation and direct participation can contribute to the promotion of EDI in companies. In Table 12.3 the different ways in which employee participation can affect organizational innovativeness and EDI are listed.

Table 12.3 Possible effects of employee participation on innovation processes

Direct Participation	Indirect Participation
Insight and commitment to business goals	Guidance for employees during processes of change
Autonomy to make suggestions and improvements	Conflict arbitration
Enhancement of knowledge flows	Feedback opportunity for management
Enrichment of management decisions	Driver and defender of innovations (if positive effects achieved on the goals of employee participation)
Culture of commitment and support	

Returning to our theoretical framework of industrial relations, we observe that the research mostly focuses on the *form* of the participation of employees in companies (direct vs. indirect). The content or the extent of the participation is rarely included in the analysis. Whether the employees are only allowed to discuss the everyday management of the company (operational participation) or can influence the general policy (strategic participation) is never included in the research. However, to fully exploit the impact of employee representation in (strategic) management decisions, these dimensions should be included, for example the attitudes of the management towards employee participation (Jirjahn and Smith, 2006). Further research should thus try to go beyond the analysis of the mere 'form' of the employee representation and investigate much more thoroughly the 'roles' – a set of connected behaviours, rights and obligations as conceptualized by actors in a social situation – this representation plays in developing forms of direct participation and innovative work behaviour. Mixed method methodologies combining both qualitative and quantitative approaches are to be developed.

Collective bargaining and innovation

Next to workplace participation, *collective bargaining* is the second key (institutional) feature by which the employment relation is governed. Here, the representatives of employees and employers negotiate different, mainly economic aspects of the formal employment relation, namely rules protecting the employment and wage trends. As wages are the primary motivator for employees to accept an employment and job security is essential for the overall well-being of employees, we can suspect an effect of both factors on employee innovative work behaviour. Although a rich literature exists about the type of collective bargaining and economic performance (Nadel, 2006), only rarely has the link been made with innovation and innovative work behaviour in particular. In the existing literature we can find only debates about the issue of labour flexibility and about the issue of wage moderation. Both topics relate to the 'output' of collective bargaining and not its process–practice. Linking these issues – labour flexibility and high-wage policies – to the question of innovation performance is anyhow still in its infancy. A brief overview is given in the following section.

The rules of engagement: hiring and firing

An important aspect of labour regulation is settlements concerning how employers can hire and fire their employees. Economists of the OECD related the 'strictness' of this type of regulation to innovative performance of countries; they concluded that strict rules are negatively related or unrelated to innovation, depending on the sector and degree of coordination of the labour relations (Bassanini and Ernst, 2002; Nicoletti *et al.*, 2001).

This research was criticized for several reasons. Their dependent and independent indicators are far from optimal. They reduce innovation to patent applications and regulation to an oversimplifying index, which is based only on the legal rules in a country. Firm-level research of Storey *et al.* (2002) adds to this critique, finding that flexible contracts were rarely introduced as a part of a plan to promote innovation; furthermore, the employees who were directly involved in innovative activities were extremely unlikely to have flexible employment relations. This is primarily because a reduced labour mobility positively affects the innovative capacity of employees, as the levels of *tacit knowledge* (which is a path-dependent form of knowledge that emerges from prior experimentation and learning) of the employees will increase. Second, it will increase the level of *commitment* of the employees. As employees have higher levels of job security, they will be more willing to engage in riskful, innovative activities for their company. Research does indeed show that workers with higher levels of job security (permanent workers) have higher levels of commitment (Jacobsen, 2000; Reisel *et al.*, 2010). Commitment is further linked to making suggestions (Parker, 2000) and organizational citizenship behaviour (van Dick *et al.*, 2008; Lavelle *et al.*, 2009; Meyer *et al.*, 2002). Moreover, employees with so-called 'contingent' contracts, who are easy to dismiss, tend to show fewer OCBs (Van Dyne and Ang, 1998). Finally, a combination of experimental and survey research by Probst *et al.* (2007) showed that job insecurity is related to poor creativity. Other research, on the other hand, found that job insecurity was positively related to OCBs, as employees tried to work harder and better to obtain more stable contracts (Feather and Rauter, 2004).

In short, the feeling of job insecurity – which is highly dependent on the objective employment status of the employee (Klandermans *et al.*, 2010) – is generally negatively related to EDI, although some opposite effects might occur. Some research finds negative links, but other research finds that contingent workers have an extra motivation to perform optimally in an attempt to increase their job security. Another strand of this literature stresses the knowledge spill-overs which an innovative economy needs. These spill-overs can only be organized if labour flexibility or job mobility is available. However, increased labour market flexibility is in these theoretical considerations rather simply defined as a determinant of knowledge flexibility. Creativity deserves flexible thinking, so it also needs flexible contractual arrangements. Again, company industrial relations and other context variables should be taken into account in further research in order to have a better grasp of the nature of this relation.

Wages and innovation

The second outcome of collective bargaining schemes is a regulated wage evolution as representatives of employees and employers come together to discuss the wage increases. A vast amount of literature exists on the relation between different types of wage bargaining and economic performance of companies and nations, but, to our knowledge, only very limited research has been conducted on the link between wage bargaining and innovation at the national level. In these mainly econometric studies no theoretical or empirical link is made with the concept and practice of EDI. We can, therefore, only speculate about the effects of wage trends on innovation and EDI. The Dutch economists Kleinknecht *et al.* (2005; 2006) and Van Schaik (2004) argue that wage moderation will be detrimental to innovation as it leads to lower investments in innovation, a slowdown of the process of 'creative destruction' and lower stimulation of demand-driven innovation. Moreover, research by Pieroni and Pompei (2008) found that wage increases over time were positively related to innovation, both for blue-collar and for white-collar workers. The efficiency wage theory develops the relation between wage and employee behaviour in more detail. The theory states that, in order to motivate employees, firms should pay above market average wages. Hereby, employees will be loyal, motivated and committed to the organization. Research indeed found that efficiency wages were positively linked to employee effort (Goldsmith, Veum, and Darity, 2000), but research linking efficiency wages to EDI is absent. In sum, further research on these questions is needed. The efficiency wage theory can serve as a good starting point here.

Conclusion and discussion

Industrial relations matter, not only generally, but also when trying to promote EDI. This can be presented as a theoretical premise, because industrial relations is about the governance of the employment relationship, which connects this 'yes' or 'no' innovating employee to his/her employer. The literature review we have presented here shows, however, a general lack of academic research linking aspects of industrial relations to innovative work behaviour of employees. Nevertheless, we can conclude that forms of employee representation in (strategic) managerial decision-making can foster EDI if embedded in positive, cooperative industrial relations. A quality employee representation, working in a trustful, cooperative relationship with the employer, can, furthermore, be positively related to direct participation, which in turn is found to be directly and strongly related to innovative employee behaviour. The effect of the principal outcomes of collective bargaining (employment protection rules and negotiated wage evolutions) is largely unknown. Very few studies have focused on these topics, although they are central to the political and societal debate. This area has enormous potential for valuable research. A more integrative approach should be considered. Specifically, the interplay between the IR climate and the outcomes of collective bargaining and their effect on EDI and employee behaviour should be further researched.

References

Acs, Z.J. and Audretsch, D.B. (1987). 'Innovation, market structure, and firm size'. *The Review of Economics and Statistics*, 69(4), 567–574.
Addison, J.T. and Belfield, C.R. (2003). 'Union voice'. *IZA Discussion Paper, 862*.
Addison, J.T. and Schnabel, C. (1996). 'German works councils, profits and innovation'. *KYKLOS*, 49(4), 555–583.
Addison, J., Kraft, K. and, Wagner, J. (1993). 'German works councils and firm performance'. In *Employee Representation: Alternatives and Future Directions*. Industrial Relations Research Association.
Addison, J.T., Schnabel, C. and, Wagner, J. (2001). 'Works councils in Germany: their effects on establishment performance'. *Oxford Economic Papers*, 53(4), 659–694.
Bassanini, A. and Ernst, E. (2002). 'Labour market institutions, product market regulation, and innovation' (p. 316). *OECD Economics Department Working Papers*.
Black, S.E., and Lynch, L.M. (2004). 'What's driving the new economy?: the benefits of workplace innovation'. *The Economic Journal*, 114, 97–116.
Block, R.N., Berg, P. and Belman, D. (2004). 'The economic dimension of the employment relationship'. In Coyle-Shapiro, J., Shore, L., Taylor, S. and Tetrick, L. (Eds) *The Employment Relationship: Examining Psychological and Contextual Perspectives* (pp. 94–117). Oxford University Press.
Blundell, R., Griffith, R. and Van Reenen, J. (1999). 'Market share, market value and innovation in a panel of British Manufacturing firms'. *Review of Economic Studies*, 66, 529–554.
Blyton, P., Dastmalchian, A. and Adamson, R. (1987). 'Developing the concept of industrial relations climate'. *Journal of Industrial Relations*, 29(2), 207–216.
Bogler, R. and Somech, A. (2005). 'Organizational citizenship behavior in school: how does it relate to participation in decision making?' *Journal of Educational Administration*, 43(5), 420–438.
Cappelli, P. and Rogovsky, N. (1998). 'Employee involvement and organizational citizenship: implications for labor law reform and "lean production"'. *Industrial and Labor Relations Review*, 51(4), 633–653.
Chen, Z.X. and Aryee, S. (2007). 'Delegation and employee work outcomes: an examination of the cultural context of mediating processes in China'. *Academy of Management Journal*, 50(1), 226.
Cooke, W.N. (1994). 'Employee participation programs, group-based incentives, and company performance: a union-nonunion comparison'. *Industrial and Labor Relations Review*, 47(4), 594–609.
Dhondt, S. and Vaas, F. (1996). *Innovatie en arbeid: Een onderzoek naar de synergie tussen kwaliteit van de arbeid en het innovatievermogen van bedrijven*. Den Haag: VUGA.
Dilger, A. (2002). *Ökonomik betrieblicher Mitbestimmung*. München: Rainer Hampp Verlag.
Feather, N.T. and Rauter, K.A. (2004). 'Organizational citizenship behaviours in relation to job status, job insecurity, organizational commitment and identification, job satisfaction and work values'. *Journal of Occupational and Organizational Psychology*, 77(1), 81–94.
FitzRoy, F.R. and Kraft, K. (1990). 'Innovation, rent-sharing and the organization of labour in the federal republic of Germany'. *Small Business Economics*, 2(2), 95–103.
Freeman, R. and Medoff, J. (1984). *What Do Unions Do?* New York: Basic Books.
Geroski, P.A. (1990). 'Innovation, Technological Opportunity, and Market Structure'. *Oxford Economic Papers*, New Series, 42(3), 586–602.

Goldsmith, A.H., Veum, J.R. and Darity, J. (2000). 'Working hard for the money? efficiency wages and worker effort'. *Journal of Economic Psychology*, 21(4), 351–385.
Gollan, P.J. and Davis, E.M. (2001). 'Employee involvement and organisational change: the diffusion of high involvement management in Australian workplaces'. In *Models of Employee Participation in a Changing Global Environment* (pp. 56–80). Aldershot: Ashgate.
Guthrie, J.P., Spell, C.S. and Nyamori, R.O. (2002). 'Correlates and consequences of high involvement work practices: the role of competitive strategy'. *The International Journal of Human Resource Management*, 13(1), 183.
Han, T.-S., Chiang, H.-H. and Chang, A. (2010). 'Employee participation in decision making, psychological ownership and knowledge sharing: mediating role of organizational commitment in Taiwanese high-tech organizations'. *The International Journal of Human Resource Management*, 21(12), 2218.
Hirsch, B.T. and Link, A.N. (1987). 'Labor union effects on innovative activity'. *Journal of Labor Research*, 8(4), 323–332.
Høyrup, S. (2010). 'Employee-driven innovation and workplace learning: basic concepts, approaches and themes'. *Transfer: European Review of Labour and Research*, 16(2), 143–154.
Hübler, O. (2003). 'Zum Einfluss des Betriebsrates in mittelgrossen Unternehmen auf Investitionen, Löhne, Produktivität und Renten–Empirische Befunde'. In Goldschmidt, N. (Ed.) *WunderbareWirtschaftsWelt – Die New Economy und ihre Herausforderungen* (pp. 77–94). Baden-Baden: Nomos.
Hyman, R. (2000). 'Social dialogue in western Europe: the "state of the art".' *Social Dialogue Papers* No. 1 (p. 39). Geneva, Switzerland: International Labour Organization.
Iverson, R.D. (1996). 'Employee acceptance of organizational change: the role of organizational commitment'. *International Journal of Human Resource Management*, 7(1), 122–149.
Jacobsen, D.I. (2000). 'Managing increased part-time: does part-time work imply part-time commitment?' *Managing Service Quality*, 10(3), 187–201.
Jirjahn, U. and Smith, S.C. (2006). 'What factors lead management to support or oppose employee participation—with and without works councils? hypotheses and evidence from Germany'. *Industrial Relations*, 45(4), 650–680.
Kaufman, B. (2004). *Theoretical Perspectives on Work and the Employment Relationship*. Industrial Relations Research Association Series. Champaign: Industrial Relations Research Association.
Kivimaki, M., Lansisalmi, H., Elovainio, M., Heikkila, A., Lindstrom, K., Harisalo, R., Sipila, K. and Puolimatka, L. (2000). 'Communication as a determinant of organizational innovation'. *R&D Management*, 30(1), 33–42.
Klandermans, B., Hesselink, J.K. and van Vuuren, T. (2010). 'Employment status and job insecurity: on the subjective appraisal of an objective status'. *Economic and Industrial Democracy*, 31(4), 557–577.
Kleinknecht, A. and Naastepad, C.W.M. (2005). 'The Netherlands: failure of a neoclassical policy agenda'. *European Planning Studies*, 13(8), 1193–1203.
Kleinknecht, A., Oostendorp, R.M., Pradhan, M.P. and Naastepad, C.W.M. (2006). 'Flexible labour, firm performance and the dutch job creation miracle'. *International Review of Applied Economics*, 20(2), 171–187.
Laursen, K. (2000). 'The importance of sectoral differences in the application of new HRM practices for innovation performance'. *DRUID Working Papers*, 01(11).

Laursen, K. and Foss, N.J. (2003). 'New human resource management practices, complementarities and the impact on innovation performance'. *Cambridge Journal of Economics*, 27(2), 243–263.

Lavelle, J.J., Brockner, J., Konovsky, M.A., Price, K.H., Henley, A.B., Taneja, A. and Vinekar, V. (2009). 'Commitment, procedural fairness, and organizational citizenship behavior: a multifoci analysis'. *Journal of Organizational Behavior*, 30(3), 337–357.

Lay, G. (1997). 'Neue Produktionskonzepte und Beschäftigung'. *Mitteilungen aus der Produktionsinnovationserhebung* No. 8 (p. 16). Karlsruhe, Germany: Fraunhofer ISI.

Machin, S. and Wadhwani, S. (1991). 'The effects of unions on investment and innovation: evidence from WIRS'. *The Economic Journal*, 101, 324–330.

Menezes-Filho, N. and Van Reenen, J. (2003). 'Unions and innovation: a survey of the theory and empirical evidence'. *International Handbook of Trade Unions* (p. 555). Cheltenham: Edward Elgar.

Meyer, J.P., Stanley, D.J., Herscovitch, L. and Topolnytsky, L. (2002). 'Affective, continuance, and normative commitment to the organization: a meta-analysis of antecedents, correlates, and consequences'. *Journal of Vocational Behavior*, 61(1), 20–52.

Michie, J. and Sheehan, M. (2003). 'Labour market deregulation, "flexibility" and innovation'. *Cambridge Journal of Economics*, 27(1), 123–143.

Nadel, H. (2006). 'Industrial relations and economic performance: an overview of research results'. *Industrial Relations in Europe 2006*. Brussels: European Commission.

Nicoletti, G., Bassanini, A., Ernst, E., Jean, S., Santiago, P. and Swaim, P. (2001). 'Product and labour markets interactions in OECD countries'. *OECD Economics Department Working Papers*, 312, 110.

OECD (1999). *Managing National Innovation Systems*. Paris: OECD.

Parker, S. (2000). 'From passive to proactive motivation: the importance of flexible role orientations and role breadth self-efficacy'. *Applied Psychology*, 49(3), 447–469.

Pieroni, L. and Pompei, F. (2008). 'Evaluating innovation and labour market relationships: the case of Italy'. *Cambridge Journal of Economics*, 32(2), 325–347.

Probst, T.M., Stewart, S.M., Gruys, M.L. and Tierney, B.W. (2007). 'Productivity, counterproductivity and creativity: the ups and downs of job insecurity'. *Journal of Occupational and Organizational Psychology*, 80(3), 479–497.

Reisel, W.D., Probst, T.M., Chia, S.-L., Maloles, C.M. and König, C.J. (2010). 'The effects of job insecurity on job satisfaction, organizational citizenship behavior, deviant behavior, and negative emotions of employees'. *International Studies of Management and Organisation*, 40(1), 74–91.

Roche, W.K. and Geary, J.F. (2002). 'Advocates, critics and union involvement in workplace partnership: Irish airports'. *British Journal of Industrial Relations*, 40(4), 659–688.

Rogers, M. (1999). 'Innovation in Australian workplaces: an empirical analysis using AWIRS 1990 and 1995'. *Melbourne Institute Working Papers* No. 3 (p. 28). Melbourne: The University of Melbourne.

Schnabel, C. and Wagner, J. (1992). 'Unions and innovative activity in Germany'. *Journal of Labor Research*, 13(4), 393–406.

Sisson, K. (1993). 'In search of HRM'. *British Journal of Industrial Relations*, 31(2), 201–210.

Sisson, K. (2008). 'Putting the record straight: industrial relations and the employment relationship'. *Warwick Papers in Industrial Relations* No. 88 (p. 57). Coventry: University of Warwick.

Storey, J., Quintas, P., Taylor, P. and Fowle, W. (2002). 'Flexible employment contracts and their implications for product and process innovation'. *The International Journal of Human Resource Management*, 13(1), 1–18.

Telljohann, V. (2010). 'Employee-driven innovation in the context of Italian industrial relations: the case of a public hospital'. *Transfer: European Review of Labour and Research*, 16(2), 227–241.

Therrien, P. and Leonard, A. (2003). 'Empowering employees: a route to innovation'. *The Evolving Workplace Series* No. 8 (p. 57). Ottawa: Statistics Canada.

Van Dick, R., van Knippenberg, D., Kerschreiter, R., Hertel, G. and Wieseke, J. (2008). 'Interactive effects of work group and organizational identification on job satisfaction and extra-role behavior'. *Journal of Vocational Behavior*, 72, 388–399.

Van Dyne, L. and Ang, S. (1998). 'Organizational citizenship behavior of contingent workers in Singapore'. *Academy of Management Journal*, 41(6), 692–703.

Van Gyes, G. (2003). *Industrial relations as a key to strengthening innovation in Europe*. Innovation Papers (no. 6). Luxemburg: Directorate-General for Enterprise, European Communities.

Van Gyes, G. (2010). 'Workplace social dialogue'. *European Company Survey 2009* Luxemburg: Eurofound.

Van Schalk, A. (2004). 'Loonmatiging gunstig voor economische groei?' *Economisch Statistische Berichten*, 91(4498), 534–536.

VanYperen, N.W., Berg, A. and Willering, M.C. (1999). 'Towards a better understanding of the link between participation in decision-making and organizational citizenship behaviour: a multilevel analysis'. *Journal of Occupational and Organizational Psychology*, 72(3), 377–392.

West, M.A. and Farr, J.L. (1990). *Innovation and Creativity at Work*. Oxford, UK: John Wiley and Sons Ltd.

Wigboldus, J.E., Looise, J.K., Nijhof, A. *et al.* (2008). 'Understanding the effects of works councils on organizational performance. a theoretical model and results from initial case studies from the Netherlands'. *Management Revue. The International Review of Management Studies*, 19(4), 307–323.

Index

Aasen, T. M. 57
active labour market policy 179
Actor Network Theory 44–45, 50, 54
ad hoc teams 6, 171, 175
adaptive leaning 17, 23–25, 28–29, 30–31, 45
adhocracy 13
affordances 16–18, 27, 87, 90, 92, 101–103
agency 16, 17, 87, 92, 94–95, 97, 99, 102–106
Alter, N. 34, 44, 51
Amabile, T. M. 110, 113, 122, 123, 128, 129, 144, 213, 214
Amundsen, O. 57
arena for interpretation 26
authority 6, 40, 47, 63, 85, 86, 171, 173, 176, 231
autonomous decision making 202
autonomy 3, 6, 13, 14, 17, 25, 26–29, 40–42, 44, 59, 65, 77, 171, 238

basic skills 149, 150, 151, 153, 154, 155
Bessant, J. 58, 59, 79, 110, 149, 157, 189, 198
Billett, S. 15–18, 22, 27, 28, 92
Bonnafous-Boucher, M. 34
Botkin, J. 23
bottom-up process 8, 9, 79, 127, 131, 139, 140, 145, 171
Boud, D. 6, 15, 18, 19, 22, 77, 89, 103
bounded agency 94, 95, 99, 104
Brandi, U. 127
break through 11, 42, 149
bridging ideas 11
building communities 11, 42, 149

capacity to innovate 57–58, 108, 191
CERI/sti 5, 12, 13, 14, 15
chaos 25, 28, 29
Chesbrough, H. 26, 59, 109, 167, 169, 170
climate 31, 41, 59, 65, 71, 110, 113, 121–123, 201, 202, 206, 230, 233, 236, 241
co-construction 47, 81

collaborative communities 167, 172–173
collaborative networks 170
collaborative team 168, 173, 175
collaborative team practices 178
collective bargaining 230–232, 239–241
collective cultures 134
community of practice 135
complex practices 10, 123, 134, 152
complexity theory 18, 27
concept of innovation 5, 16, 109
conflict 22, 26, 54, 85, 122, 138, 179, 226
constrains of EDI 220, 223, 225
convergent processes 4, 24–27, 31
cooperate solidarity 154
cooperation 9, 64, 71, 193, 236, 238
corporate cultures 133
creative leadership 122
creativity 13, 19, 24–25, 30, 65, 87, 108, 114, 120, 123, 129, 142–144, 188, 240
criss-crossing teams 171, 174–175, 178
critical reflection 27, 30
cross company innovations 202
cultural characteristics 57, 58, 62, 64, 67, 68
cultural manifestation 133
cultural pattern of meaning 132
cultural practices 5, 16, 77, 100, 144

Darsø, L. 108
decision making 43, 44, 65, 66, 96, 97, 110, 121, 159, 162, 201, 202, 205, 232, 236, 241
design driven innovation 34
developmental learning 22, 23, 24, 26, 27, 43
Diamond of innovation 109, 111, 113–116, 119, 123
direct participation 7–8, 230–231, 236, 238–239, 241
discretion 13, 15, 17, 25, 94, 96–97, 99, 101–102, 104, 171
discretionary learning 13–15
distributed innovation 169–170

Index

divergent processes 4, 25–27, 29–31
drivers of innovation 6, 80, 127, 128, 131, 178, 187, 191
Drucker, P. 10, 11, 58, 79, 80, 88, 109, 112, 123, 187, 191, 202
DUI-mode 15

ecology of learning 150, 156–157
EDI 35–37, 41–45, 47–48, 50–54, 62–71, 127–128, 135, 140–146, 171, 173, 178, 180–181, 187–191, 199–200, 204–206, 214, 217–218, 220, 222–225, 230, 232–233, 236, 238, 240–241. *See also* Employee-driven Innovation
EDI apparatuses 45
EDI management apparatuses 45
EDI practices 57–58, 62, 64, 66–71
EDI tools 68
Ellström, P.-E. 17, 22–29, 45, 131
employee apparatuses 45
employee behavior 230, 232–233, 236, 238, 241
employee involvement 59, 63, 67–68, 71, 150, 159, 193, 197, 200–201, 203
employee learning 3, 13, 14, 43, 54
employee participation 9, 34–35, 50, 57, 71, 198, 230
employee representation 231–233, 236, 238–239, 241. *See also* shop steward; union representatives
Employee-driven Innovation 1, 3–4, 6–12, 14, 16, 18–20, 22, 24, 26, 28–29, 31, 34–35, 37–51, 57–58, 60, 66, 70, 77–81, 83, 92–95, 97, 101–102, 105–106, 129–131, 136, 139–142, 144, 149, 157, 159–160, 162, 165, 167–169, 179, 185, 198, 211–212, 216, 226–227. *See also* EDI
employee-entrepreneurs 44, 46
employee-led innovation 16, 18, 89, 92–95, 104
employees' innovative potential 6, 143
employees' role 52, 201, 204
empowerment 68, 187, 198, 206, 212
enact 5, 10, 17, 18, 77, 78, 80, 81, 87, 89, 92–95, 97–100, 104
enmeshment of practices 18, 81
entrepreneurial approaches 83, 84

entrepreneurship 37–39, 41, 129, 130, 142, 230
European Innovation Scoreboard 34
European Working Conditions Survey 13
Evans, K. 15, 149, 150, 151, 152, 154, 157, 158
everyday cultural practices 77
everyday practice 3, 89, 137, 153
everyday practice learning 80
experience based learning 80, 131, 135, 141
exploratory learning 28

Fees, W. 185
Fenwick, T. xi, 18, 80, 89
first order EDI 9
formalized approach to EDI 50

geographically distributed teams 181
global networks 165, 167, 170
global teams 168, 181
governance 49, 62, 82, 170, 178, 231, 241
Gressgård, L. J. 57
Gyes, G.V. 230

Hansen, K. 57
Hargadon, A. 11, 112–113, 121, 172, 214
Hasse, C. 127
high-involvement innovation 189, 198
high-performance work organization 178
Høyrup, S. 3, 15, 25, 34, 50, 51, 53, 54, 79, 114, 131, 145, 187, 188, 189, 198, 211, 212, 232
hybrid innovation 42, 43

ignorance 111–118
improvisational learning 19, 21–22, 29
improvising 20–22
improvising as learning 22
incremental 13, 16–17, 37, 39, 42–43, 79, 149, 157, 181, 198
indirect participation 230–231, 238
individual learning 3, 16, 21, 23, 190
individual tactics 153
industrial relations 12, 152, 178, 230, 231–232, 233, 236, 237, 238, 239, 240, 241

informal learning 6, 149, 150, 154, 157, 160, 162
informal networks 19, 20, 25
informal organization 20, 25
innovating in global networks 167
innovation abilities/ skills 178, 186
innovation capability 185, 187, 189, 191, 197, 205, 206
innovation capability measurement instrument 192
innovation competency 108–111, 116, 117, 121, 123, 124
innovation culture 133, 145
innovation leadership roles 114
innovation leadership skills 186
innovation measurement 234, 237
innovation model 117
innovation processes 3–5, 7, 9–12, 15–17, 19, 22–25, 30, 31, 35, 39, 45, 52, 54, 57–60, 70, 90, 109–111, 114–117, 123, 128, 130–132, 143, 144, 202, 203, 206, 211, 238
innovation strategy 90, 191, 205
innovations at work 92–95, 101
innovative behavior 13, 40, 47, 129, 236
innovative learning 4, 17, 20, 22–31, 51, 168–170
innovative practices 79, 94, 98, 100, 101, 131, 169, 170
innovative processes 5, 8–10, 17, 29, 31, 46, 111
institutional flexibility 162
integrated development practices 78, 89
intra-innovative competency 111
intrinsic motivation 122
involvement 6, 9, 10, 57, 58, 59, 63, 64, 65, 66, 67, 68, 71, 80, 88, 130, 150, 156, 159, 173, 175, 200, 201, 203, 206, 232, 233, 236, 238
involvement of unions 156

Kanter, R. 51, 58, 201, 214
Kersh, N. 149
Kesting, P. 7, 8, 9, 187, 189, 191, 198
key innovation dimension 199–200
knowing 116, 117, 118, 120, 122, 123

knowledge sharing 170, 173–174, 176, 236
Kristensen, P. H., 167

lateral collaborative networks 170
Lave, J. 94, 135, 214
Leadership 109, 110, 111, 113, 114–116, 119, 121, 122, 123, 133, 156, 186, 189, 191, 198
leadership roles 110, 114–116, 123
Lean 13–14
learning and innovation 3–4, 11–13, 15–16, 19–22, 24–25, 30–31
learning organizations 12–13, 171, 178, 180
learning processes 4, 10, 12, 15–17
learning space 117 124
Lempiälä, T. 211
linear model of innovation 128
linking learning and innovation 15, 112
LO (Confederation of Danish Trade Unions) 7, 8, 9, 34, 130
loosely coupled network structure 20
Lorenz, E. and Valeyre, A. 13, 178
Lotz, M. 167
low skilled employees 134

maintenance learning 23
management 8–11, 14, 19–20, 26–27, 35–40, 43–47, 53–54, 65–68, 70, 82–86, 121, 123, 129–146, 170–172, 175–177, 185–193, 196–206, 221–227
managing employee-driven innovation 38, 47, 130
manifestations of EDI 218, 220, 222, 225
measurement instrument 192
Miner, A. S. 19–22
motivation 35, 47, 51, 59, 102, 122–123, 131, 154, 157–159, 161, 187

national business system 179
nature of work 77, 78
negotiate 85, 104, 106, 141, 143, 178–180, 219, 220, 231, 232, 239, 241
networks 12, 19–20, 112, 167, 169–170, 213
Nieuwenhuis, L. 46

non intimidating spaces 157
Nonaka, I. and Takeuchi, H. 109, 112, 120–121
non-R&D-based innovative forms 168–169

occupational practices 16, 17, 92, 94, 100, 105
OECD 5, 12, 238
open innovation 26, 34, 50, 54, 59, 109
organizational boundaries 104, 109, 174
organizational citizenship 113
organizational conditions 4, 24, 29, 31, 71, 115
organizational culture 12, 29, 62, 68, 127, 128–139, 141–146
organizational development 211, 218
organizational entrepreneurship 37–39, 41–43
organizational expectations 52, 85
organizational innovation 31, 36, 77, 131, 140, 236
organizational learning 3, 13, 21–23, 135, 190–192
organizational moves 71, 167, 169, 171, 173, 175, 178, 179
organizational norms 44, 45, 54
organizational practices 8, 18, 23, 78, 81–83, 85–90, 180
organizational routines 25, 28–29, 202
organizational strategies 47, 153
outcome of innovation 17, 59, 71

paradoxes of EDI 35
participation 7–9, 34, 35, 37, 39, 50, 57, 67, 70, 71, 79, 81, 101, 130, 149, 150, 189, 198, 214, 218, 230–234, 236–239, 241
participatory innovations 40
phases of innovation 110
practical intelligibility 18, 81, 82, 87
practice based innovation 22, 45, 168, 178
practice based learning 127, 131–145
practice based workplace culture 132
practice theory 78, 80
preject 30, 31, 114, 115, 144
presentational knowing 116
Price, O. M. 5, 6, 7, 10, 17, 18, 19, 30, 77

problem solving 3, 13, 14, 19, 23, 25–27, 29, 98, 121, 214
process innovation 22, 145, 202, 234, 237
product innovation 233–235
project 10, 13, 28, 30, 31, 35, 37, 39, 40, 42, 44, 45, 47–53, 65, 66, 70
propositional knowing 116

radical innovations 39, 42, 43, 44, 45, 52, 189
R&D Communities 211–213, 215–217, 226
recombinant innovation 112
reconstruction of work practice 10, 16
Redien-Collot, R. 34
reflection 20, 27, 28, 29, 30, 54, 117, 118, 120, 198, 214
relationship building 118
remaking of job 8, 18, 83
remaking of the work practice 15–17, 19, 21, 31, 89
reproductive learning 23, 26, 43, 51
resources of innovation 5, 6, 7
restricted participation 218
risk taking 27, 43, 122
roles 52, 57, 58, 62, 63, 65, 67–70, 80, 87, 103, 104, 110, 114–116, 123, 157, 159, 170, 171, 174–178, 180, 188, 205, 239
routines 6, 19, 20, 23–25, 28, 29, 66, 67, 131, 171, 174, 175, 192, 202, 220

Schatzki, T. R. 81
Scheeres, H. 77
Schumpeter, J. A. 5, 42, 79, 109, 127, 128, 137, 169
scientist-entrepreneur 217–221, 230, 236
second order EDI 10
self management 24, 27
self-directed learning 28
serendipity 37, 52
shop steward 175, 178, 179, 231, 232. *See also* employee representation; union representatives
skills crises 151
skills for life 151
social dialogue 232
social ecology 152

social innovation 10, 123
socio-innovative competency 110, 121
socio-personal account 16
socio-personal account of innovation 92–95
source of innovation 6, 34, 44, 49, 50, 54, 59
Spiegelaere, S. D. 230
strategic participation 238
sub cultures 133
successful innovation 46, 62, 111, 114, 191, 192, 198, 199, 237
Sundbo, J. 19, 20, 25, 79, 212

tacit knowledge 117, 120, 189, 240
Taherizadeh, A. H. 185
task complexity 13, 14
Taylorism 13, 14, 15
team-based work-organizing 169
technology-driven innovation 34
Teglborg, A.-C. 9, 10, 34, 35, 36, 45, 46, 48, 54, 79, 189
tension 18, 25, 26, 35, 43, 44, 53, 54, 88, 94, 211
third order EDI 10
Tidd, J. 58, 59, 188, 189, 191, 192, 198, 203
top-down processes 9, 10
Total Innovation Capability 196, 197, 206
trade unions 7, 58, 60, 130, 152, 231, 236
traditional organizational form 14
trust 52, 88, 113, 114, 117–122, 202

union representatives 57, 60, 62, 64, 65, 68–71, 159, 175–177.
See also employee representation; shop steward
union support 156
user-driven innovation 34, 59, 169

Van de Ven 58, 110, 113, 117
variety 17, 22, 25, 26, 28, 29, 50, 66, 114, 151, 153, 154, 156, 160, 161
Viala, C. 34
virtual learning 160

wages and innovation 240
Waite, E. 149, 152, 155, 157, 158
welfare state 179
Wenger, E. 135, 214
work communities 211–214
work organization 3, 12–15, 19, 22, 24, 26, 30, 31, 47, 149, 151–155, 161, 204, 205
work practice 5, 8, 9, 30, 77, 82, 93, 95, 96, 100, 101, 140, 149, 161, 167, 169, 171, 172, 174, 176, 178, 180, 213, 214, 224
working environment 43
workplace cultures 134, 140
workplace innovations 98
workplace learning 3, 6, 10, 15, 22–23, 25, 28, 30–31, 47, 149, 151–155, 161, 211–212
works council 231, 233, 235, 237

Yli-Kauhaluoma, S. 211